Y0-EOA-812

WITHDRAWN

PROJECTIONS ONTO TRANSLATION-INVARIANT SUBSPACES OF $L^p(G)$

Haskell P. Rosenthal

University of Minnesota

and

Université de Paris, Faculté des Sciences d'Orsay

Contents

This work was supported in part by NSF Grant GP-3617, an NSF cooperative graduate fellowship, and an NSF-NATO postdoctoral fellowship, and is based on a doctoral dissertation written under the supervision of Professor Karel de Leeuw at Stanford University.

I would like to express my deep appreciation to Professor de Leeuw for his patience, advice, and constant encouragement. I would also like to thank Professor Yitzhak Katznelson and Professor Isadore I. Hirschman for stimulating conversations that I had with them on topics related to the contents of this Memoir.

Received by the editors March 22, 1965 and, in revised form, December 27,1965

INTRODUCTION

Let A be a closed translation-invariant subspace of $L^p(G)$, $1 \leq p \leq \infty$, G a locally compact abelian group. (The letter "A" always denotes such a Banach space in what follows, "group" always refers to a locally compact abelian group.)

We shall be concerned with the following two questions :

1. Under what circumstances is there a bounded linear projection from $L^p(G)$ onto A? (Alternatively, when is A complemented in $L^p(G)$?)

2. Under what circumstances is A linearly homeomorphic to Hilbert space?

One of the main results concerning the first question is Theorem 1.4, where we obain necessary conditions for the existence of a bounded linear projection from $L^1(G)$ onto A, where A is an invariant subspace of $L^1(G)$ and G is arbitrary.

Precisely, if $f \in L^1(G)$, let $\hat{f}$ be the Fourier transform of f, and let hull A=hA= $\{\tau$ in the dual group of G : $\hat{f}(\tau) = 0$, all $f \in A\}$.
We demonstrate in Theorem 1.4 that there is no bounded projection onto A unless hA belongs to the coset-ring of the discrete dual of G (i.e., hA is a member of the smallest ring of sets containing some coset of the dual of G).

As corollaries we obtain that if G is the real line, A is uncomplemented unless hA is discrete (Theorem 1.6); if G is the integer group A is complemented if and only if hA is a finite subset of the circle (Corollary 1.7).

The machinery developed to prove Theorem 1.4 together with some properties of injective spaces enables us to demonstrate that the space of continuous analytic functions on the circle is not linearly homeomorphic to the space of all continuous functions on any compact Hausdorff space (see the Remark following the proof of 1.8) ; the generalization of this result to

QA
3
A57
no.63

an arbitrary compact group with ordered dual is established in Theorem 1.8 and Corollary 1.9. This same machinery also enables us to show that quotients of the group algebra of a particular group are not usually linearly homeomorphic to any group algebra (Corollary 1.10). (The group algebra of G refers to $L^1(G)$ under convolution).

The crucial step in the proofs of the above results in Theorem 1.1, where we show that if A is a weak*-closed translation-invariant subspace of $L^\infty(G)$, and if there is a bounded projection from $L^\infty(G)$ onto A, then there is a bounded projection onto A that commutes with translations.

It follows from this (Corollary 1.3) that if G is compact, and if A is weak* closed, then there is a bounded projection from $L^\infty(G)$ onto A if and only if there is a weak* continuous projection onto A that commutes with translations.

We give an example in Section I of an invariant subspace A of infinite codimension which is complemented in L^1 of the real line even though there is no projection onto A that commutes with translations.

However, it is impossible to find such a subspace A for $L^1(G)$ if G is compact, or for $L^p(G)$, G being arbitrary, $1 < p < \infty$. (The analogue of this result for non-commutative locally compact amenable groups holds as well; see the remark following Lemma 3.1.)

Indeed, it has been established by Walter Rudin in [16], that if G is a compact group, and A is an invariant subspace of $L^p(G)$, then if A is complemented in $L^p(G)$, A is invariantly complemented. I.e. there is a projection from $L^p(G)$ onto A that commutes with translations.

In Lemma 3.1, we extend Rudin's result to arbitrary abelian groups, for $1 < p < \infty$.

It is well known that a bounded operator on $L^p(G)$ that commutes with translations is given by a multiplier. We thus obtain that if $1 < p \leq 2$, and A is complemented in $L^p(G)$, then there is a set E measurable in the dual of G, such that

$$A = \{f \in L^p(G) : \hat{f}(\tau) = 0 \text{ a.e.}, \tau \notin E\}.$$

Moreover, for each $f \in L^p(G)$, $\chi_E \hat{f} = \hat{g}$ for some $g \in A$. I.e., χ_E multiplies $L^p(G)$. (χ_E denotes the <u>characteristic function</u> of the set E.)

We show in Theorem 3.2 that if $u \in L^p(\mathbb{R}^n)$, $1 < p < 2$, $u \neq 0$, then there exists an open set S such that $\chi_S \hat{u}$ is not the (a.e.) transform of any L^p function. It then follows from the preceding paragraph that

$$A_p = \{f \in L^p(\mathbb{R}^n) : \hat{f}(x) = 0 \text{ almost all } x \notin S\}$$

is uncomplemented in $L^p(\mathbb{R}^n)$ (see Corollary 3.5).

In Corollary 3.4 we show that if $u \in L^1 \cap L^2$, $u \neq 0$, then there exists an open set $S \subset \mathbb{R}^n$ such that $\chi_S \hat{u}$ is not the (a.e.) transform of any L^p function, all $1 < p < 2$. We then conclude in Corollary 3.5 that for this S the A_p defined above is uncomplemented in $L^p(\mathbb{R}^n)$, all $1 < p < 2$.

The last result in Section III is Theorem 3.8, where we show that if A is a non-zero closed translation-invariant subspace of $L^p(\mathbb{R}^n)$, $1 < p < 2$, then there is a closed translation-invariant subspace of A that is uncomplemented in A.

In Theorem 2.2 we obtain necessary and sufficient conditions that a closed translation-invariant subspace be complemented in L^p of the circle, $1 < p < \infty$.

This result enables us to construct for each $p \neq 2, 1 < p < \infty$, uncomplemented subspaces of L^p (Corollary 2.3).

We first deal with the second question for compact G. We know that if A is linearly homeomorphic to Hilbert space, then every closed invariant subspace of A is complemented in A ; indeed, in Hilbert space we merely take the subspace's orthogonal complement. In Theorem 2.9 we show that the converse of this is true for $p < \infty$. Precisely, let A be a closed translation-invariant subspace of $L^p(G)$, (G compact), and suppose that A satisfies the following condition :

Every closed translation-invariant subspace of A is complemented in A.

Then A is linearly homeomorphic to Hilbert space provided $1 \leq p < \infty$. However, if $p = \infty$ and A is infinite dimensional and weak* closed, then A is not even reflexive ; in fact, we show in Theorem 2.7 that this condition guarantees that every member of A has an absolutely convergent Fourier series.

Returning again to the case $1 \leq p < \infty$, we also demonstrate in Theorem 2.9 the following two necessary and sufficient conditions that A be linearly homeomorphic to Hilbert space :

(i) $A \subset L^2(G)$ and L^2 and L^p norms on A are equivalent.

(ii) Every function obtained by convolving a member of A with a member of $L^q(G)$ has an absolutely convergent Fourier series $(\frac{1}{p} + \frac{1}{q} = 1)$. We also obtain necessary and sufficient conditions in Theorem 2.11 that A be both complemented in $L^p(G)$ and linearly homeomorphic to Hilbert space ; the relationship between such A's and the lacunary series dealt with in [17] and [6] is discussed.

Finally, one of the lemmas used in obtaining the above results (Lemma 2.4) enables us to show in Corollary 2.6 that no quotient of the group algebra of any group can be linearly homeomorphic to Hilbert space unless it is finite dimensional.

The last result of the second section shows that the aformentioned results dealing with Question 2 are quite properly stated for compact groups; they have no content for non-compact groups.

For we show in Corollary 2.13 that if G is a non-compact group, there are no non-zero invariant Hilbert subspaces of $L^p(G)$, ($1 \leq p < \infty$, $p \neq 2$) (i.e., if A is a non-zero closed translation-invariant subspace of $L^p(G)$, then A is not linearly homeomorphic to Hilbert space).

Moreover, we show in the same result that when $p = 1$, a non-zero invariant subspace of $L^1(G)$ is not even reflexive.

Corollary 2.13 is a fairly immediate consequence of Theorem 2.12, where we show that A has a subspace linearly homeomorphic to ℓ^p.

DEFINITIONS, NOTATION, AND SOME STANDARD FACTS

We presuppose the knowledge of standard results in harmonic analysis and the theory of Banach spaces. Results for which no explicit reference is given may be found in [4] and [15].

For the most part, the notations and definitions that are used below are standard ; those that deal with Banach space theory may be found in [4]; those that deal with harmonic analysis may be found in [15].

(i) General Banach Spaces

Throughout (i) A and B will denote complex Banach spaces.

A and B are said to be isomorphic if A and B are linearly homeomorphic.

Fact 1 : This is equivalent, by the open mapping theorem, to the existence of a bounded linear operator $T : A \to B$, with T one-one and onto.

A and B are said to be isometric if there is a norm-preserving linear map from A onto B.

If $A \subset B$, A is said to be complemented in B with norm $\leq K$, if there exists a bounded linear projection P from B onto A, with norm $\leq K$. I.e., $P^2 = P$, range $P = A$, $\|P\| \leq K$.

A is said to be complemented in B, if A is complemented in B with norm $\leq K$ for some K.

Fact 2 : By the closed graph theorem the latter is equivalent to the existence of a closed subspace $C \subset B$, with $A \oplus C = B$ (see p. 480 of [4]). I.e., $A \cap C = \{0\}$, and if $b \in B$, there exist $a \in A$, $c \in C$, with $a + c = b$.

If A cannot be complemented in B, i.e. if there is no bounded linear projection onto A, we say that A is uncomplemented in B.

Fact 3 : If $A \subset B$ and $\pi : B \to B/A$ is the natural homeomorphism from B onto B/A, then B/A is a Banach space under the norm " $|||\cdot|||$ " defined on B/A by

$$|||x||| = \inf_{y \in A} \|y-x\|,$$

where $\|\cdot\|$ denotes the norm on B.

Fact 4 : If A and C are closed subspaces of B, and if $A \oplus C = B$, then B/C is isomorphic to A.

A^* denotes the adjoint of A ; i.e., A^* is the Banach space consisting of all bounded linear functionals from A into the complex numbers.

If $B \subset A$, $B^\perp$, the annihilator of B, is defined by

$$B^\perp = \{f \in A^* : f(b) = 0, \text{ all } b \in B\}.$$

Fact 5 : If $B \subset A$, then B^* is isometric to $A^*/B^\perp$, and $(A/B)^*$ is isometric to $B^\perp$.

Fact 6 : If A is isomorphic to B, then A^* is isomorphic to B^*.

The weak* topology on A^* is defined by the following basis for open neighborhoods of the origin.

Let $a_1, \ldots, a_n \in A$ and $\epsilon > 0$.

$$U_{a_1,\ldots,a_n;\epsilon} = \{f \in A^* : |f(a_i)| < \epsilon, \ i = 1,\ldots, n\}.$$

We then have that if $\{f_\alpha\}_{\alpha \in D}$ is a net in A^* (D a directed set), then $f_\alpha \to f$ (weak*) if and only if $f_\alpha(a) \to f(a)$, each $a \in A$.

If C is a weak* closed subspace of A^*, C is said to be weak* complemented in A^* if there exists a weak* closed subspace $D \subset A^*$ with $C \oplus D = A^*$. C is said to be norm-complemented if C is complemented in the Banach space A^*;

i.e., if there exists a bounded linear projection from A^* onto C.

<u>Fact</u> 7 : If $A \subset B$, then A is complemented in B if and only if $A^{\perp}$ is weak* complemented in B^*.

<u>Fact</u> 8 : Let $A \subset B$ and $T : B \longrightarrow B$ a bounded linear operator. Then T is a projection onto A if and only if both (i) and (ii) hold.

(i) Range $T \subset A$.

(ii) If $f \in A$, then $Tf = f$.

(ii) <u>Harmonic Analysis</u>

$\mathbb{R}$ denotes the additive group of real numbers, $\mathbb{C}$ the additive group of complex numbers, $\mathbb{Z}$ the additive group of integers, and $\mathbb{T}$ the multiplicative circle group. ($\mathbb{T} = \{z : z \in \mathbb{C}, |z| = 1\}$.)

If H is a group, H^n denotes the n-fold sum $H + \dots + H$.

<u>Throughout this paper</u>, G <u>denotes a locally compact abelian group</u> ; Γ <u>denotes the dual of</u> G (all groups will be assumed to be locally compact abelian, unless otherwise specified).

Γ thus consists of all continuous homomorphisms $\tau : G \longrightarrow \mathbb{T}$. If $\tau \in \Gamma$, the function $\tau : G \longrightarrow \mathbb{T}$ will also be denoted by

$$x \longrightarrow (\tau, x) ;$$

or simply by $(\tau, \cdot)$.

For example, if $G = \mathbb{R}^n$, then $\Gamma = \mathbb{R}^n$. Thus if $x \in \mathbb{R}^n$, then $(x, \cdot)$ denotes the function

$$y \longrightarrow (x,y) = \exp(i \sum_{j=1}^{n} y_j x_j) ,$$

where $x = (x_1, \dots, x_n)$ and $y = (y_1, \dots, y_n)$.

<u>The coset-ring of</u> G <u>is the smallest ring of sets on</u> G <u>containing the open cosets of</u> G. I.e., the coset-ring of G equals $\bigcap\{F : F$ is a ring of sets on G, and every open coset of G is a member of $F\}$.

(A ring of sets on G is simply a family of subsets of G closed under complementation and finite unions).

<u>A trigonometric polynomial on the group</u> G is a continuous complex-valued function f of the form $f(\cdot) = \sum_{i=1}^{n} a_i(\tau_i, \cdot)$, where $a_i \in \mathbb{C}$ and $\tau_i \in \Gamma$, $1 \leq i \leq n$.

The space of <u>continuous almost-periodic functions</u> on G is defined to be the closure of the space of trigonometric polynomials in supremum norm, i.e. all f on G such that there exists a sequence P_n of trigonometric polynomials on G with $P_n \to f$ uniformly.

C(G) denotes the space of <u>uniformly continuous complex-valued functions</u> defined on G, under the supremum norm. Note that C(G) does not denote the space of bounded continuous functions on G unless every such function is actually uniformly continuous. (For example, it is well known that if G is compact, then every continuous function on G is uniformly continuous.)

$C_0(G)$ consists of all members of C(G) vanishing at infinity, i.e. $f \in C_0(G)$ if and only if f is continuous, and given $\epsilon > 0$, there is a compact set K so that $|f(x)| < \epsilon$ all $x \in G \sim K$.

If H is a group, then H_d denotes the group H under the discrete topology.

The <u>Bohr compactification</u> of G, denoted by $\tilde{G}$, is defined as the dual of Γ_d. We shall use the following facts about $\tilde{G}$ (see pp. 30-32 of [15]).

1. The elements of the dual of $\tilde{G}$ and the dual of G are the same.

2. G may be regarded as being densely imbedded in $\tilde{G}$, i.e. there is a map $\beta : G \longrightarrow \tilde{G}$; β is a continuous 1-1 homomorphism of G onto a dense subgroup of $\tilde{G}$.

3. Using 1. and 2., $C(\tilde{G})$ may be identified with the continuous almost-periodic functions on G.

Where there can be no confusion, $\int_G f(x)dx$ denotes integration with respect to some Haar measure on G.

The simpler notations $\int f$, $\int_G f$, or $\int_G fdx$ may also be used.

The term "invariance of Haar measure" refers to both inversion and translation invariance of Haar measure. That is, if $f \in C(G)$, f compactly supported, then for all $\tau \in G$,

$$\int_G f(x+\tau)dx = \int_G f(x)dx = \int_G f(-x)dx.$$

Haar measure is normalized for compact non-discrete G by total mass $= 1$; for discrete G by point-mass $= 1$; for $\mathbb{R}^n$ by Haar measure $=$ Lebesgue measure.

For fixed G , $L^p(G)$ is the Banach space of equivalence classes of pth power Haar-integrable complex-valued functions $(1 \leq p \leq \infty)$. The norm $\|\cdot\|_{L^p(G)}$ is given by

$$\|f\|_{L^p(G)} = \left(\int_G |f|^p dx\right)^{\frac{1}{p}} ,$$

and will be denoted as $\|f\|_p$ when there is no possibility of confusion. ($\|f\|_\infty$ = essential supremum norm, of course.)

The adjoint space of $L^p(G)$ may be identified with $L^q(G)$, where $1 \leq p < \infty$ and $\frac{1}{p} + \frac{1}{q} = 1$ (see E10, p. 268 of [15].) <u>Throughout this paper,</u>

p and q will always denote two numbers satisfying $\frac{1}{p} + \frac{1}{q} = 1$.

We shall take the identification of $(L^p(G))^*$ with $L^q(G)$ in the following way : If $g \in L^q(G)$, then the representation of g as a linear functional on $L^p(G)$ is given by

$$f \longrightarrow \int_G f(-x)g(x)dx, \text{ all } f \in L^p(G).$$

If G is discrete, we note that $L^p(G)$ is the space of all functions $f : G \to \mathbb{C}$ such that $\sum_{g\in G} |f(g)|^p < \infty$. In this case, $L^p(G)$ will also be denoted by $\ell^p(G)$. If $G = \mathbb{Z}$, the simpler notations ℓ^p and ℓ will be used for $\ell^p(\mathbb{Z})$ and $\ell^1(\mathbb{Z})$, respectively.

A finite, regular, complex-valued Borel measure on G will simply be called a finite measure on G.

By the Riesz representation theorem, $C_0(G)^*$ may be identified with the space of finite measures on G.

If μ is a finite measure on G , we will take the representation of as a linear functional on $C_0(G)$ by

$$f \longrightarrow \int_G f(-x)d\mu(x) , \text{ all } f \in C_0(G) .$$

If f and g are a pair of Borel-measurable functions on G , the convolution of f and g , denoted $f*g$, is defined by $f*g(y) = \int_G f(y-x)g(x)dx$ whenever $\int_G f(y-x)g(x)\, dx < \infty$ (for properties of convolution, see [15], p. 4).

If $f \in L^1(G)$, then the Fourier transform of f , denoted by $\hat{f}$, belongs to $C_0(\Gamma)$ and is defined by

$$\hat{f}(\gamma) = \int_G f(x)(\gamma,-x)dx , \text{ all } \gamma \in \Gamma .$$

More generally, if μ is a finite measure on G , then the Fourier-Stieltjes transform of μ , denoted by $\hat{\mu}$, belongs to $C(\Gamma)$ and is defined by

$$\hat{\mu}(\tau) = \int_G (\tau,-x)d\mu(x) \text{ , all } \tau \in \Gamma .$$

$A(G)$ denotes the space of Fourier-transforms of $L^1(\Gamma)$. I.e., $f \in A(G)$ if and only if there is a $g \in L^1(\Gamma)$ such that

$$f(x) = \hat{g}(x) = \int_\Gamma (\tau,-x)g(\tau)d\tau \text{ for all } x \in G .$$

We then define

$$\|f\|_{A(G)} = \|g\|_{L^1(\Gamma)} .$$

(The definition makes sense by the uniqueness of Fourier transform.)

In the case in which G is compact, $A(G)$ will be called the space of absolutely convergent Fourier series on G. Since Γ is then discrete, $A(G)$ is naturally isometric to $\ell^1(\Gamma)$. We have that $f \in A(G)$ if and only if $\sum_{\tau\in\Gamma} |\hat{f}(\tau)| < \infty$; moreover, if this is the case, $f(x) = \sum_{\tau\in\Gamma} \hat{f}(\tau)(\tau,x)$, all $x \in G$.

The following definitions may not be in such common usage as the preceding ones.

If $x \in G$, R_x denotes the operation of translating functions by x. That is, if f is a function defined on G, then

$$(R_x f)(y) = f(x+y) \text{ , all } y \in G .$$

Now, fix G, and let $\mathcal{B}$ denote one of the Banach spaces $L^p(G)$, $1 \leq p < \infty$, or $C(G)$.

1. A is said to be an invariant subspace of $\mathcal{B}$, if A is closed and if $R_x f \in A$ for all $x \in G$ and $f \in A$.

2. If A is an invariant subspace of $\mathcal{B}$, and if $T : A \longrightarrow A$, T is said to be invariant if T is a bounded linear operator with $(R_{-x}TR_x)f = Tf$ for all $x \in G$ and $f \in A$; that is, if T commutes with all translations.

3. If A is an invariant subspace of $\mathcal{B}$, and if C is an invariant subspace of A, C is said to be invariantly complemented in A if there is an invariant projection P from A onto C. From Fact 2 of (i) and the definition of invariance, this is equivalent to the existence of an invariant subspace D, $D \subset A$, with $C \oplus D = A$.

4. If G is compact, let A be an invariant subspace of $\mathcal{B}$. Otherwise, let A be an invariant subspace of $L^1(G)$.

The hull of A, denoted by hA, is defined by $hA = \{\tau \in \Gamma : \hat{f}(\tau) = 0$, all $\tau \in A\}$. Thus hA is the set of characters of G in A. Moreover, hA is a closed subset of Γ. For if $f \in \mathcal{B}$, $\hat{f}$ is continuous, whence $Z_f = \{\tau \in \Gamma : \hat{f}(\tau) = 0$ is a closed subset of Γ. Hence $hA = \bigcap_{f \in A} Z_f$ is a closed set.

SECTION I

PROJECTIONS ONTO TRANSLATION - INVARIANT SUBSPACES OF $L^1(G)$

The main result of the first section is Theorem 1.4, which states that if A is an invariant subspace of $L^1(G)$, G arbitrary, and $S = hA$, then A cannot be complemented in $L^1(G)$ unless S belongs to the coset ring of Γ_d.

The proof of this is accomplished by showing first that if a weak* closed invariant subspace of $L^\infty(G)$ is norm-complemented in $L^\infty(G)$, then it is invariantly norm-complemented (Theorem 1.1). It is then demonstrated that an invariant projection on $L^\infty(G)$ gives rise to an idempotent measure on the Bohr compactification of G (Lemma 1.2). Finally, Cohen's theorem characterizing the idempotent measures (see [11], or [15], p. 60) is applied to complete the proof of the main result.

As an immediate corollary of Theorem 1.1 and Lemma 1.2, one obtains that if G is compact, and A is an invariant subspace of $L^1(G)$, then, if $A^\perp$ is norm-complemented in $L^\infty(G)$, A is invariantly complemented in $L^1(G)$ (Corollary 1.3).

Corollary 1.5 applies Theorem 1.4 to show that one cannot usually extend restrictions of Fourier transforms in a continuous linear fashion.

The next two results apply Theorem 1.4 to give necessary conditions that invariant subspaces of $L^1(\mathbb{R})$ and ℓ be complemented. For ℓ the problem is completely solved : we show in Corollary 1.7 that a non-zero invariant subspace of ℓ is complemented if and only if it is of finite codimension. We show in Theorem 1.6 that if A is a non-trivial complemented subspace of $L^1(R)$, then hA differs by a finite set from a finite union of stretched cosets of the integers.

For $L^1(R)$ the problem is not completely solved, since we do not know wheter or not every A of this form is complemented. However, if $hA = \mathbb{Z}$, then A is complemented (see the remark following Theorem 1.4).

The remaining results of this section require some properties of injective spaces ; the definition of these spaces and the properties that are used are mentioned explicitly before Theorem 1.8 ; in this theorem we show that if G is a compact group with ordered dual, and if

$$C_A = \{f \in C(G) : \hat{f}(\tau) = 0 \text{ , all } \tau < 0\} ,$$

then C_A^{**} is not injective. In Corollary 1.9 we show that C_A cannot be linearly homeomorphic to the space of continuous functions on any compact Hausdorff space, thus obtaining as a special case that the space of continuous analytic functions on the circle is not isomorphic to $C(\mathbb{T})$.

The final result, Corollary 1.10, shows that if G is arbitrary, A an invariant subspace of $L^1(G)$, and if hA is not in the coset-ring of Γ_d, then $L^1(G)/A$ is not isomorphic to $L^1(X, \mathcal{S}, \mu)$, for any measure space $(X, \mathcal{S}, \mu)$. In particular, this means that quotients of the group algebra are not usually even linearly homeomorphic to other group algebras.

Theorem 1.1. *Let $B \subset A \subset L^\infty(G)$, G arbitrary, B a weak* closed invariant subspace, A a norm-closed invariant subspace.*

Suppose $P : A \longrightarrow B$ is a projection, bounded in supremum norm. Then there exists a projection $T : A \longrightarrow B$, T invariant and bounded in supremum norm.

Proof. Let M be an invariant mean on $\ell^\infty(G)$. That is, if f is a bounded function on G, $M(R_g f) = Mf$, all $g \in G$; $M(1) = 1$,

$|M(f)| \leq \|f\|_\infty$, M linear.*

Let $\langle\, ,\, \rangle$ denote the pairing between L^∞ and L^1. So if $f \in L^\infty$, $h \in L^1$, $\langle f,h\rangle = \int_G f(x)h(-x)dx$. Now fix $f \in L^\infty$. Then $h \longrightarrow M_x \langle R_{-x}PR_x f,h\rangle$ defines a bounded linear functional on $L^1(G)$, of norm $\leq \|P\|\,\|f\|$. Letting Tf be the unique member of L^∞ representing the functional, T appears as a bounded linear operator $T : A \longrightarrow L^\infty$, with $\|T\| \leq \|P\|$.

Next, T is a projection onto B. It suffices to show that range $T \subset B$, and that if $f \in B$, $Tf = f$.

Range $T \subset B$: We know that $B^{\perp\perp} = B$, since B is weak* closed. So $B = \{g : \langle g,h\rangle = 0$, all $h \in B^\perp\}$. Suppose $h \in B^\perp$. Then $\langle R_{-x}PR_x f,h\rangle = 0$, each $x \in G$, $f \in A$; since $R_{-x}PR_x f \in B$. Thus $M_x \langle R_{-x}PR_x f,h\rangle = \langle Tf,h\rangle = 0$, each such h, so $Tf \in B$.

$Tf = f$, $f \in B$: If $f \in B$, $PR_x f = R_x f$; whence $\langle Tf,h\rangle = M_x \langle R_{-x}R_x f,h\rangle = M_x \langle f,h\rangle = \langle f,h\rangle$, all $h \in L^1$, so $Tf = f$.

Finally, T is invariant. For fix $f \in A$, $h \in L^1$, $y \in G$.

$$\langle R_y Tf,h\rangle = \langle Tf,R_y h\rangle .$$

So

$$\langle R_y Tf,h\rangle = M_x \langle R_{-x}PR_x f,R_y h\rangle .$$

But

$$\langle R_{-x}PR_x f,R_y h\rangle = \langle R_y R_{-x}PR_x f,h\rangle , \text{ for all } x \in G .$$

So

---

* The existence of such a mean follows from the Markov-Kakutani fixed-point theorem (p. 456 of [4]) ; or see [19].

$$\langle R_y Tf,h\rangle = M_x \langle R_y R_{-x} PR_x f,h\rangle$$

$$= M_x \langle R_y R_{-(x+y)} PR_{(x+y)} f,h\rangle \text{ (by invariance of } M \text{ and } A)$$

$$= \langle TR_y f,h\rangle .$$

Thus

$$R_y T = TR_y , \text{ all } y \in G .$$

(de Leeuw used an argument similar to this to obtain invariant projections on $C_0(G)$. See Theorem 4.1 of [7].)

Remark. Even if A and B are weak* closed and P is weak* continuous, T need not be weak* continuous. See the remark at the end of Theorem 1.4.

In the next lemma we make use of the Bohr compactification $\hat{G}$ of a group G. As was pointed out in the Introduction, the dual of $\hat{G}$ is simply Γ endowed with the discrete topology ; and the almost-periodic continuous functions on G may be identified with $C(\hat{G})$ (see [15], pp. 30-32 for properties of $\hat{G}$).

Lemma 1.2. Let G be arbitrary, A an invariant subspace of $L^1(G)$, and let $T : L^\infty(G) \longrightarrow A^\perp$ be an invariant, norm-bounded projection onto A . Then there is a finite measure μ on the Bohr compactification G , so that $\hat{\mu}$ is the characteristic function of hA, such that $Tf = f*\mu$ for each continuous almost-periodic function f .

Proof. The invariance of T implies that the class of uniformly continuous functions is invariant under T . For if $f \in C(G)$,

$$\|R_y Tf-Tf\|_\infty = \|T(R_y f-f)\|_\infty \leq \|T\| \, \|R_y f-f\|_\infty \to 0 \text{ as } y \to 0 \text{ in } G .$$

Thus Tf is (the equivalence class of) a uniformly continuous function, by an argument identical to that given in the first paragraph, p. 132, of [7].

So $f \longrightarrow (Tf)(0)$ defines a bounded linear functional on the almost-periodic continuous functions, i.e., on $C(\tilde{G})$. So there is a measure μ on $\tilde{G}$, with $(Tf)(0) = \int_{\tilde{G}} f(-y)d\mu(y)$. But for $x \in G$, $(Tf)(x) = (R_x Tf)(0) = TR_x f(0) = \int_{\tilde{G}} R_x f(-y)d\mu(y) = f*\mu(x)$. So $Tf = f*\mu$, all $f \in C(\tilde{G})$.

Now, since T is a projection, $T^2 = T$, whence $\mu*\mu = \mu$. I.e., μ is idempotent. Thus $(\hat{\mu})^2 = \hat{\mu}$, so $\hat{\mu}(\tau) = 1$ or 0, for all $\tau \in \Gamma_d$. Now

$$T(\tau, .) = \mu*(\tau, .) = \hat{\mu}(\tau)(\tau,.)$$

for all $\tau \in \Gamma_d$. Also, $\tau \in hA$ if and only if $(\tau,.) \in A^\perp$. So, if $\tau \in hA$, then $T(\tau,.) = (\tau,.)$ hence $\hat{\mu}(\tau) = 1$. Otherwise, $T(\tau,.) \neq (\tau,.)$, so $\hat{\mu}(\tau)$ cannot be 1 and hence must be 0.

As an immediate consequence of Theorem 1.1 and Lemma 1.2, since $G = \tilde{G}$ for G compact, we obtain

Corollary 1.3. *Let G be a compact group, and A an invariant subspace of $L^1(G)$. Then if $A^\perp$ is norm-complemented in $L^\infty(G)$, there is a finite idempotent measure μ on G, with $\hat{\mu}(\tau) = 1$ if and only if $\tau \in hA$, all $\tau \in \Gamma$.*

Remark. Since G is a compact group, the trigonometric polynomials in A are dense in A (see Spectral Synthesis Theorem of Section II). It then follows that

$$T : L^1(G) \longrightarrow L^1(G)$$

defined by

$$Tf = f - f*\mu, \text{ all } f \in L^1(G),$$

is a projection from $L^1(G)$ onto A. Hence if B is a weak* closed invariant subspace of $L^\infty(G)$, then B is norm-complemented in $L^\infty(G)$ if and only if B is weak* invariantly complemented in $L^\infty(G)$.

For take A invariant in $L^1(G)$, with $B = A^{\perp}$. If B is norm-complemented, we have thus demonstrated the existence of an invariant projection $T : L^1(G) \to A$. Then $I - T^*$ is a weak* continuous invariant projection onto B. If B is weak* complemented, B is norm-complemented, this being true in the adjoint of any Banach space.

Theorem 1.4. *Let* G *be arbitrary,* A *an invariant subspace of* $L^1(G)$, *and* $S = hA$. *Then* A *is uncomplemented in* $L^1(G)$ *unless* S *belongs to the coset-ring of* Γ_d (*i.e.,* Γ *with the discrete topology*).

Proof. Suppose $P : L^1(G) \to A$ is a projection onto A. Letting P^* be the adjoint of P, we have that $I - P^*$ is a projection from $L^\infty(G)$ onto $A^{\perp}$. By Theorem 1.1 there is an invariant projection $T : L^\infty(G) \longrightarrow A^{\perp}$, onto $A^{\perp}$. By Lemma 1.2 there is an idempotent measure μ on the Bohr compactification $\bar{G}$, so that $\hat{\mu}(\tau) = 1$ if and only if $\tau \in hA$. Finally, by Cohen's theorem ([11], or p. 60 of [15]), hA must belong to the coset-ring of the dual of G, namely Γ_d. Q.E.D.

Remark. A may be complemented without being invariantly complemented, even if A is of infinite codimension. We shall give here an example of such an A; it then follows that the invariant projection $T : L^\infty(G) \longrightarrow A^{\perp}$ that one obtains from Theorem 1.1 cannot be weak* continuous, since otherwise A would be invariantly complemented.

Let $G = \mathbb{R}$, and let

$$A = \{ f \in L^1(\mathbb{R}) : \hat{f}(n) = 0 \text{ , all integers } n \}.$$

A is complemented in $L^1(\mathbb{R})$ by the functions supported on the interval $[0,2\pi]$. Indeed, if we define $U : L^1(\mathbb{R}) \longrightarrow L^1(\mathbb{R})$ by

$$(Uf)(x) = \sum_{n=-\infty}^{\infty} f(x+2\pi n) \text{ for } x \in [0,2\pi],$$

$$(Uf)(x) = 0 \text{ for } x \notin [0,2\pi],$$

then $(Uf)(x)$ is well defined for almost all x, $Uf \in L^1(\mathbb{R})$, and $I - U$ is a projection from $L^1(\mathbb{R})$ onto A.

However, suppose A were invariantly complemented. By 3.8.4, p. 75, of [15] the invariant projection P from $L^1(\mathbb{R})$ onto A would be of the form $Pf = f*\mu$ for some finite measure μ. We would then have $\hat{\mu}(n) = 0$ for all integers n, $\hat{\mu}(x) = 1$ for all real non-integers x. This is impossible, since then $\hat{\mu}$ would not be continuous on $\mathbb{R}$.

The next corollary is, in a certain sense, simply a restatement of Theorem 1.4. If S is a closed subset of Γ, define

$$B_S = \{g : g = \hat{f}|S \text{ , some } f \in L^1(G)\}.$$

If we define

$$D_S = \{f \in L^1(G) : \hat{f}(s) = 0 \text{ , all } s \in S\},$$

then B_S is naturally algebraically isomorphic to $L^1(G)/D_S$. So we simply endow B_S with the norm given to $L^1(G)/D_S$.

Now the following corollary states that if S is not in the coset-ring of Γ_d, one cannot continuously linearly extend elements of B_S to Fourier transforms defined on all of Γ, i.e. to elements of $A(G)$.

Corollary 1.5. *If S is not in the coset-ring of Γ_d, there is no bounded linear map $T : B_S \longrightarrow A(G)$ with $Tg|S = g$, all $g \in B_S$.*

Proof. It is easily demonstrated that such a map T exists if and only if D_S is complemented in $L^1(G)$.

The next two results follow, essentially, from Theorem 1.4 and the fact that if A is an invariant subspace of $L^1(G)$, G arbitrary, then hA is a closed subset of Γ . For if $f \in L^1(G)$, then

$$Z_f = \{\tau \in \Gamma : \hat{f}(\tau) = 0\}$$

is a closed subset of Γ , since $\hat{f}$ is continuous. Hence $hA = \bigcap_{f \in A} Z_f$ is a closed set. Thus, if A is complemented, then hA must be a closed subset of Γ, and at the same time belong to the coset ring of Γ_d . So Theorem 1.6 and Corollary 1.7 are obtained by explicating the sets $E \subset \Gamma$ such that E is both a closed subset of Γ and a member of the coset-ring of Γ_d .

Theorem 1.6. *Let A be a non-trivial invariant subspace of $L^1(\mathbb{R})$. If A is complemented in $L^1(\mathbb{R})$, then there is a finite set of real numbers F , and real numbers $\tau_1, \ldots, \tau_n$; $\beta_1, \ldots, \beta_n$, such that*

$$hA \cup F = \bigcup_{i=1}^{n} (\tau_i \mathbb{Z} + \beta_i) .$$

Proof. We are assuming that $A \neq 0$, $A \neq L^1(\mathbb{R})$, and that A is complemented in $L^1(\mathbb{R})$. Then hA is a proper non-empty closed subset of $\mathbb{R}$, and a member of the coset-ring of $\mathbb{R}_d$ by Theorem 1.4.

Let $\mathcal{R}$ denote the coset-ring of $\mathbb{R}_d$ ($\mathcal{R}$ thus consists of the smallest ring of sets containing cosets of arbitrary subgroups of $\mathbb{R}$). Let $E \in \mathcal{R}$, E a closed non-empty subset of $\mathbb{R}$, $E \neq \mathbb{R}$.

We shall show here that E must be discrete. The proof that E differs by a finite set from a finite union of stretched cosets of the integers (i.e., that E satisfies the same properties as the set hA mentioned in the statement of Theorem 1.6), requires a more detailed analysis of $\mathcal{R}$; it is given in A0 of the Appendix. (A0 plus the discreteness of E thus completes the proof of Theorem 1.6.)

First, E must be of measure zero. For χ_E is measurable and also the Stieltjes transform of a finite measure on the Bohr compactification of $\mathbb{R}$; hence χ_E is equal almost everywhere to a continuous function, as is well known (see [5]).

We show next that arbitrarily large bounded subsets of E must belong to $\mathcal{R}$. Precisely, that for any a, $b \in \mathbb{R} \sim E$, $[a, b] \cap E$ belongs to $\mathcal{R}$. Since $\mathbb{R} \sim E$ is open, we may choose a $\delta > 0$, so that

$$[a-\delta, a] \cap E = [b, b+\delta] \cap E = \emptyset .$$

Now choose a function φ supported in $[a-\delta, b+\delta]$ with $\varphi = 1$ on $[a,b] \cap E$, such that a translate of φ is positive definite.

Then

$$\chi_{[a,b] \cap E} = \varphi \cdot \chi_E$$

so $\chi_{[a,b] \cap E}$ is the transform of an idempotent measure on the Bohr compactification of $\mathbb{R}$, whence $[a,b] \cap E \in \mathcal{R}$.

To show that E is discrete, we need merely show that a bounded member of $\mathcal{R}$ must be finite. But any infinite member of $\mathcal{R}$ contains (with the exception of finitely many points) an infinite coset (see the lemma in AO of the Appendix): since cosets of infinite subgroups of $\mathbb{R}$ are never bounded, the discreteness of E is established.

Remark. We do not know whether or not the converse to Theorem 1.6 holds. As was pointed out in the Remark following Theorem 1.4, if $hA = \mathbb{Z}$, then A is complemented in $L^1(\mathbb{R})$. However, suppose α is an irrational number; let

$$B = \{f \in L^1(\mathbb{R}) : \hat{f}(n) = \hat{f}(\alpha n) = 0, \text{ all } n \in \mathbb{Z}\},$$

i.e.,

$$hB = \mathbb{Z} \cup \alpha\mathbb{Z} .$$

We do not know whether or not B is complemented in $L^1(\mathbb{R})$.

<u>Corollary</u> 1.7. <u>Let</u> A <u>be a non-zero invariant subspace of</u> ℓ . <u>Then</u> A <u>is complemented in</u> ℓ <u>if and only if</u> A <u>is of finite codimension, i.e. if and only if</u> hA <u>is finite</u>.

<u>Proof</u>. Assume that A is a proper subspace; i.e., $A \neq 0$, $A \neq \ell$. Then hA is a closed, non-empty proper subset of $\mathbb{T}$. Now assume that A is complemented. Then by Theorem 1.4, hA is a member of the coset-ring of $\mathbb{T}_d$, the discrete circle.

Let $\varphi : \mathbb{R} \to \mathbb{T}$ be the map

$$\varphi(x) = e^{ix}.$$

φ is continuous; therefore $E = \varphi^{-1}(hA)$ is a closed subset of $\mathbb{R}$. ($E=\{x\in\mathbb{R} : \varphi(x) \in hA\}$). Moreover, since φ is a homomorphism, $\varphi^{-1}(F)$ is a coset of a subgroup of $\mathbb{R}$ if F is a coset of a subgroup of $\mathbb{T}$. Finally, since complements and finite intersections are preserved under inverse images, $\varphi^{-1}(hA) = E$ belongs to the coset-ring of $\mathbb{R}_d$.

It follows from the proof of Theorem 1.6 that E is either discrete, or $E = \mathbb{R}$ ($E \neq \emptyset$ by definition).

Now

$$\varphi(E \cap [0,2\pi]) = hA .$$

Since $A \neq 0$, $hA \neq \mathbb{T}$, so $E \neq \mathbb{R}$. Hence E is discrete, whence $E \cap [0,2\pi]$ is finite, so hA is finite; thus A is of finite codimension.

On the other hand, it is well known that for any Banach space B, a closed subspace of finite codimension is always complemented in B. Hence if A is of finite codimension, A is complemented.

Q.E.D.

Remark. As a special case of Corollary 1.7, we have that if S is a closed infinite proper subset of $\mathbb{T}$, and if

$$B = \{f \in A(\mathbb{T}) : f(x) = 0, \text{ all } x \in S\},$$

then B is uncomplemented in $A(\mathbb{T})$.

The proofs of the final results of this section require some properties of injective spaces. For the proofs of the properties mentioned below, see [10] and page 94 of [2].

Definition. A Banach space A is called injective if, given any Banach space B with $A \subset B$,

A is complemented in B.

Some properties of injective spaces are :

(1). *If A is injective and A is isomorphic to a Banach space D, then D is injective.*

(2). *Complemented subspaces of injective spaces are injective.*

(3). *If $(X, \mathcal{S}, \mu)$ is a measure space, then $(L^1(X))^*$ is injective.*

(4). *If Y is a compact Hausdorff space, then $C(Y)^{**}$ is injective.*

($C(Y)$ denotes all continuous complex-valued functions defined on Y under supremum norm; $C(Y)^{**}$ is the second-adjoint space of $C(Y)$.)

Theorem 1.8. *Let G be an infinite compact group with ordered dual (as defined in Chap. 8 of [15]). Let $C_A = \{f \in C(G) : \hat{f}(\tau) = 0,$ all $\tau < 0\}$. Then C_A^{**} is not injective.*

We first state an immediate corollary, then prove the theorem.

Corollary 1.9. *C_A is not linearly homeomorphic to $C(X)$ for any*

<u>compact Hausdorff space</u> X.

<u>Proof</u>. If C_A were isomorphic to $C(X)$, then C_A^{**} would be injective, by Properties (1) and (4) of injective spaces.

<u>Proof of Theorem</u> 1.8.

First we need some notation.

Let "$\approx$" denote the relation "is isomorphic to".

Let "C" stand for $C(G)$, "L^1" stand for $L^1(G)$.

Let B_s be the set of all $\mu \in C^*$ that are singular with respect to Haar measure on G.

Let $H_a = C_A^{\perp} \cap L^1$; $H_s = C_A^{\perp} \cap B_s$.

Now, by the Lebesgue decomposition Theorem (p. 132 of [3]),

$$C^* = L^1 \oplus B_s .$$

By the generalized F. and M. Riesz theorem on analytic measures (8.2.3, p. 200, of [15]),

$$C_A^{\perp} = H_a \oplus H_s .$$

(Note that

$$C_A^{\perp} = \{\mu : \mu \text{ a finite measure on } G, \hat{\mu}(\tau) = 0, \text{ all } \tau > 0\}.)$$

Thus

$$C_A^* \approx \frac{C^*}{C_A^{\perp}} \approx \frac{L^1 \oplus B_s}{C_A^{\perp}}$$

$$\approx \frac{L^1 \oplus B_s}{H_a \oplus H_s}$$

$$\approx \frac{L^1}{H_a} \oplus \frac{B_s}{H_s} .$$

We thus have that $\frac{L^1}{H_a}$ is isomorphic to a complemented subspace of C_A^*.

Indeed, let

$$U : \frac{L^1}{H_a} \oplus \frac{B_s}{H_s} \to C_A^* ,$$

U 1-1, onto, linear, and bounded. Then

$$U\left(\frac{L^1}{H_a}\right) \dot{\oplus} U\left(\frac{B_s}{H_s}\right) = C_A^* .$$

Thus $H_a^{\perp}$ is isomorphic to a complemented subspace of C_A^{**}. For we have that $\left(U\left(\frac{B_s}{H_s}\right)\right)^{\perp}$ is complemented in C_A^{**} by $\left(U\left(\frac{L^1}{H_a}\right)\right)^{\perp}$.

But since

$$\frac{L^1}{H_a} \approxeq U\left(\frac{L^1}{H_a}\right) ,$$

$$\left(\frac{L^1}{H_a}\right)^* \approxeq \left(U\left(\frac{L^1}{H_a}\right)\right)^* .$$

Moreover,

$$\left(\frac{L^1}{H_a}\right)^* \approxeq H_a^{\perp} .$$

Thus

$$H_a^{\perp} \approxeq U\left(\frac{L^1}{H_a}\right)^*$$

$$\approxeq \left(U\left(\frac{B_s}{H_s}\right)\right)^{\perp} .$$

Now suppose C_A^{**} were injective. Then H_a would be injective, by Properties (1) and (2) of injective spaces. Hence $H_a^{\perp}$ would be norm-complemented in $L^{\infty}(G)$. (Note that $H_a^{\perp} = \{f \in L^{\infty}(G) : \hat{f}(\tau) = 0,$

all $\tau > 0\}$.)

By Corollary 1.3 there is a finite measure μ on G ,
$\hat{\mu}(\tau) = 1$ if $\tau \in h(H_a)$, $\hat{\mu}(\tau) = 0$ otherwise. Hence by Cohen's theorem, (3.1.3, p. 60, of [15]). $h(H_a)$ is in the coset-ring of Γ . However, $h(H_a) = \{\tau \in \Gamma : \tau > 0\}$ is not in the coset-ring of Γ .

This contradiction shows that C_A^{**} is not injective.

Remark. We recently discovered that Pełczynski has independently established our Theorem 1.8 for the case in which G is the circle, and has thus obtained the following special case of our Corollary 1.9. (see [14]):

The space of all functions continuous on the closed unit disc and holomorphic in its interior, is not linearly homeomorphic to C(X) for any compact Hausdorff space X. (It is well known that this space may be identified with $C_A(\mathbb{T})$) .

Corollary 1.10. *Let G be arbitrary, and let A be an invariant subspace of $L^1(G)$. If hA is not in the coset-ring of Γ_d, $L^1(G)/A$ cannot be isomorphic to $L^1(X,\mathcal{S},\mu)$, where $(X,\mathcal{S},\mu)$ is any measure space.*

Proof. Suppose

$$L^1(G)/A \approx L^1(X,\mathcal{S},\mu) .$$

Then

$$(L^1(G)/A)^* \approx (L^1(X,\mathcal{S},\mu))^* .$$

But

$$(L^1(G)/A)^* \approx A^\perp .$$

Thus $A^\perp$ is injective, by Property (3) of injective spaces. Hence $A^\perp$ is norm-complemented in $L^\infty(G)$. Combining Theorem 1.1 and Lemma 1.2, we have that there is a finite measure μ on the Bohr compactification $\overset{\frown}{G}$, with

$$\hat{\mu}(\tau) = 1 \text{ if } \tau \in hA ,$$
$$\hat{\mu}(\tau) = 0 \text{ if } \tau \notin hA , \text{ all } \tau \in \Gamma .$$

Thus by Cohen's theorem, hA is in the coset-ring of Γ_d.

Q.E.D.

SECTION II

TRANSLATION - INVARIANT HILBERT SUBSPACES OF $L^p(G)$, G COMPACT ABELIAN

Throughout this section G denotes a compact abelian group, Γ the dual of G. $\mathcal{B}$ denotes one of the Banach spaces $L^p(G)$, $1 \leq p < \infty$, or $C(G)$. If $f \in \mathcal{B}$, the *spectrum* of f, denoted by sp f, is defined to be $\{\tau \in \Gamma : \hat{f}(\tau) \neq 0\}$. If A is an invariant subspace of $\mathcal{B}$, or a weak* closed invariant subspace of $L^\infty(G)$, the spectrum of A, denoted by sp A, is defined as $\bigcup_{f \in A} \text{sp } f$.

We will need the following known result from harmonic analysis.

Spectral Synthesis Theorem: *If A is an invariant subspace of* $\mathcal{B}$, *then*

$$A = \{g \in \mathcal{B} : \text{sp } g \subset \text{sp } A\}.$$

If A is a weak* closed subspace of $L^\infty(G)$, *then*

$$A = \{g \in L^\infty(G) ; \text{sp } g \subset \text{sp } A\}.$$

This theorem follows from 2.7 of [18].

Remark: Thus if A_1, A_2 are two invariant subspaces of $\mathcal{B}$, and sp A_1 = sp A_2, then $A_1 = A_2$. In particular, if A is an invariant subspace, and A_1 = the closure of the set of trigonometric polynomials in A, then sp A_1 = sp A ; hence, since $A_1 = A$, the trigonometric polynomials in A are dense in A.

Spectral Synthesis Corollary: *Let D and A be invariant subspaces of* $\mathcal{B}$, $D \subset A$. *Then D is invariantly complemented in A if and only if* $\sum_{\tau \in \text{spD}} \hat{f}(\tau)(\tau, \cdot)$ *is the Fourier series of a member of* $\mathcal{B}$ *for each* $f \in A$. *Moreover, an invariant projection P from A onto D satisfies the relation* $\widehat{Pf}(\tau) = \chi_{\text{spD}}(\tau)\hat{f}(\tau)$ *for all* $f \in A, \tau \in \Gamma$,

<u>and is hence unique</u>.

<u>Proof</u>: Suppose D is invariantly complemented in A : we may then write that $A = D \oplus B$, where B is invariant, B is the null space of an invariant projection P. The Spectral Synthesis Theorem now yields that

$$B = \{f \in A : \text{sp } f \cap \text{sp } D = \emptyset\} .$$

Now fix $f \in A$. Then $f = g + h$, where $g = P(f) \in A$, and $h \in B$. Then

$$\hat{f}(\tau) = \hat{g}(\tau) + \hat{h}(\tau), \text{ for all } \tau \in \Gamma .$$

If $\tau \in \text{sp } D$, $\hat{h}(\tau) = 0$, since $h \in B$; hence $\hat{g}(\tau) = \hat{f}(\tau)$. If $\tau \notin \text{sp } D$, $\hat{g}(\tau) = 0$, since $g \in D$. Thus

$$\hat{g}(\tau) = \widehat{Pf}(\tau) = \chi_{\text{spD}}(\tau)\hat{f}(\tau) , \text{ for all } \tau \in \Gamma,$$

so the Fourier series for Pf is $\sum_{\tau \in \text{spD}} \hat{f}(\tau)(\tau, .)$.

Conversely, define $P : A \to D$ by $Pf = g$ if g is the element of $\mathcal{A}$ having Fourier series $\sum_{\tau \in \text{spD}} \hat{f}(\tau)(\tau, .)$.

P is well defined by the uniqueness of Fourier transform; since $\text{sp } Pf \subset \text{sp } D$, $Pf \in D$ by the Spectral Synthesis Theorem. $P^2 = P$, so P is a projection from A onto D. It remains to show that P is bounded.

Let $f_n \to f$, $Pf_n \to g$; f_n , $f \in D$, $g \in A$. Then $\hat{f}_n \to \hat{f}$ and $\widehat{Pf}_n \to \hat{g}$ uniformly, hence pointwise. $\widehat{Pf}_n = \chi_{\text{spD}} \hat{f}_n$ by definition of P ; thus since $\hat{f}_n \to \hat{f}$ pointwise, $\chi_{\text{sp D}} \hat{f}_n \to \chi_{\text{sp D}} \hat{f}$ pointwise. So $\chi_{\text{sp D}} \hat{f} = \hat{g}$; hence $Pf = g$.

Thus by the closed graph theorem, P is a bounded operator.

Q.E.D.

We shall need some properties of Sidon sets in this section.

Definition: If $S \subset \Gamma$, S is called a Sidon set if $\sum_{\tau \in \Gamma} |\hat{f}(\tau)| < \infty$ for each $f \in C(G)$ with $\operatorname{sp} f \subset S$.

Suppose S is a Sidon set.

Define $A = \{f \in C(G) \; ; \; \operatorname{sp} f \subset S\}$.

By an easy application of the closed graph theorem, we obtain that A is isomorphic to $\ell(S)$. ($\ell(S)$ is defined as the space of all functions $g : S \to \mathbb{C}$ such that $\sum_{s \in S} |g(s)| < \infty$, under the norm

$$\|g\|_{\ell(S)} = \sum_{s \in S} |g(s)| \;).$$

Another property of S is given by Theorem 5.7.3, p. 120, of [15] :

if $f \in L^{\infty}(G)$, $\operatorname{sp} f \subset S$, then $\sum_{\tau \in \Gamma} |\hat{f}(\tau)| < \infty$.

Hence $A = \{f \in L^{\infty}(G) : \operatorname{sp} f \subset S\}$, and A is thus weak* closed. Moreover, it follows from the Spectral Synthesis Theorem that every weak* closed invariant subspace of A is norm-complemented in A. Indeed, if B is a weak* closed invariant subspace of A, then $A = B \oplus C$ where $C = \{f \in A : \operatorname{sp} f \cap \operatorname{sp} B = \emptyset\}$; C is weak* closed, so B is actually weak* complemented.

One of the results of this section is the converse of the last fact: Namely, if A is a weak* closed invariant subspace of $L^{\infty}(G)$, and if every weak* closed invariant subspace of A is norm-complemented in A, then $\operatorname{sp} A$ is a Sidon set (Theorem 2.7).

We establish also in this section an analogue of this result for $L^p(G)$. We prove in Theorem 2.9 that if $A \subset L^p(G)$, $1 \le p < \infty$, and if every invariant subspace of A is complemented in A, then A is isomorphic to Hilbert space.

The use of the Rademacher functions is essential for the proof of this result (see Lemma 2.8 and "(c) $\Longrightarrow$ (d)" of Theorem 2.9). These functions are defined as follows:

Let $r_0(t)$ be of period 1, $r_0(t) = +1$ for $0 \leq t < \frac{1}{2}$

$$r_0(t) = -1 \text{ for } \tfrac{1}{2} \leq t < 1 .$$

<u>The nth Rademacher function r_n is defined by</u>

$$r_n(t) = r_0(2^n t) , \quad 0 \leq t \leq 1 .$$

It is easily seen that the Rademacher functions form an orthonormal set in $L^2[0,1]$ (see p. 6 of [20]).

The main property of these functions is as follows:

Let the <u>Rademacher polynomials</u> be the members of the linear span of $r_0, r_1, r_2, \ldots$. I.E., f is a Rademacher polynomial if and only if there are complex constants $a_0, \ldots, a_n$ so that $f = \sum_{k=0}^{n} a_k r_k$.

<u>Theorem</u> 2.0. <u>The L^r norms on the Rademacher polynomials are all equivalent,</u> $1 \leq r < \infty$. <u>Precisely, for each $r \in [1, \infty)$, there are positive constants A_r and B_r, so that for all Radamacher polynomials</u> f,

$$A_r \|f\|_{L^2[0,1]} \leq \|f\|_{L^r[0,1]} \leq B_r \|f\|_{L^2[0,1]} .$$

(This is stated as Theorem 8.4, on p. 213 of [20].) (Note that if $f = \sum_{k=0}^{n} a_k r_k$, then

$$\|f\|_{L^2[0,1]} = \sum_{k=0}^{n} \left(|a_k|^2 \right)^{\frac{1}{2}}$$

by the orthonormality of the r_n 's.)

We first mention a theorem of Walter Rudin, which shows that with respect to invariant subspaces of $\mathcal{B}$, complemented and invariantly complemented mean the same thing. For its proof, see [16].

Theorem 2.1 [Rudin] . *Let* $D \subset A$, *where* D *and* A *are invariant subspaces of* $\mathcal{B}$. *If* D *is complemented in* A , *then* D *is invariantly complemented in* A .

The next theorem is a fairly simple consequence of the Spectral Synthesis Corollary. Its interest derives from the fact that because of Theorem 2.1 it actually characterizes the invariant subspaces of $L^p(\mathbb{T})$ that are complemented in $L^p(\mathbb{T})$ (recall that $\mathbb{T}$ denotes the circle group).

Theorem 2.2. *Let* A *be an invariant subspace of* $L^p(\mathbb{T})$, $1 < p < \infty$. *Then* A *is invariantly complemented in* $L^p(\mathbb{T})$ *if and only if*

$$\lim_{N \to \infty} \sum_{j=-N}^{N} \chi_{spA}(j)\hat{f}(j)\hat{g}(j) \text{ exists, all } f \in L^p , g \in L^q .$$

Proof. Suppose A is invariantly complemented. By the Spectral Synthesis Corollary.

$$\sum_{j=-\infty}^{\infty} \chi_{spA}(j)\hat{f}(j)e^{ijx} = \sum_{j \in spA} \hat{f}(j)e^{ijx}$$

is the Fourier series of a member of $L^p(\mathbb{T})$ for each $f \in L^p(\mathbb{T})$; denoting this member by h , we have by Theorem 6.4, p. 266, of [20] , that

$$\frac{1}{2\pi} \sum_{j=-N}^{N} \chi_{spA}(j)\hat{f}(j)e^{ijx}$$

converges to h in the L^p mean, as $N \to \infty$. Hence, for $g \in L^q$,

$$\lim_{N \to \infty} \sum_{j=-N}^{N} \chi_{spA}(j)\hat{f}(j)\hat{g}(j) = \lim_{N \to \infty} \frac{1}{2\pi} \int_0^{2\pi} \left(\sum_{j=-N}^{N} \chi_{spA}(j)\hat{f}(j)e^{ijx} \right) g(-x)dx$$

$$= \frac{1}{2\pi} \int_0^{2\pi} h(x)g(-x)dx \text{ , so the limit exists.}$$

(Here we have identified functions on the circle with 2π-periodic functions defined on the real line.)

Conversely, fix $f \in L^p(\mathbb{T})$. For each integer N, define a bounded linear functional F_N on $L^q(\mathbb{T})$ by

$$F_N(g) = \sum_{j=-N}^{N} \chi_{spA}(j)\hat{f}(j)\hat{g}(j), \text{ all } g \in L^q(\mathbb{T}) .$$

By assumption, $\lim_{N \to \infty} F_N(g)$ exists for each $g \in L^q(\mathbb{T})$. Thus, by the Banach-Steinhaus Theorem, the linear functional F defined on $L^q(\mathbb{T})$ by

$$F(g) = \lim_{N \to \infty} F_N(g)$$

is bounded. So we know there is an $h \in L^p(\mathbb{T})$ satisfying

$$F(g) = \frac{1}{2\pi} \int_0^{2\pi} g(x)h(-x)dx \text{ , all } g \in L^q(\mathbb{T}) .$$

Thus

$$\hat{h}(n) = \frac{1}{2\pi} \int_0^{2\pi} e^{inx}h(-x)dx$$

$$= F(e^{in\cdot})$$

$$= \chi_{spA}(n)\hat{f}(n) \text{ by definition of } F .$$

so $\sum_{n \in spA} \hat{f}(n)e^{in\cdot}$ is the Fourier series of an $h \in L^p(\mathbb{T})$ for each $f \in L^p(\mathbb{T})$, so A is invariantly complemented in $L^p(\mathbb{T})$ by the

Spectral Synthesis Corollary.

Corollary 2.3. *If* $1 < p < \infty$ *and* $p \neq 2$, *there exists an invariant subspace of* $L^p(\mathbb{T})$ *that is uncomplemented in* $L^p(\mathbb{T})$.

Proof. Suppose $1 < p < 2$. Choose $f \in L^p(\mathbb{T})$, $f \notin L^2(\mathbb{T})$. Then $\sum_{-\infty}^{\infty} |\hat{f}(n)|^2 = \infty$. It will be shown in Lemma 2.8 that there is a continuous g with

$$\sum_{-\infty}^{\infty} |\hat{f}(n)\hat{g}(n)| = \infty .$$

(This fact is well known anyway; for example, it is an immediate consequence of the theorem of p. 222 of [15] .)

Now choose a subset S of the integers with

$$\lim_{j\to\infty} \sum_{n=-j}^{j} \chi_S(n)\hat{f}(n)\hat{g}(n) \text{ non-existent.}$$

(To construct such an S, let

$$A_1 = \{n : \operatorname{Re} \hat{f}(n)\hat{g}(n) > 0\}$$

$$A_2 = \{n : \operatorname{Re} \hat{f}(n)\hat{g}(n) < 0\}$$

$$A_3 = \{n : \operatorname{Im} \hat{f}(n)\hat{g}(n) > 0\}$$

$$A_4 = \{n : \operatorname{Im} \hat{f}(n)\hat{g}(n) < 0\}.$$

Then either

$$\sum_{n\in A_i} \operatorname{Re} \hat{f}(n)\hat{g}(n) = + \text{ or } -\infty \text{ if } i = 1 \text{ or } 2,$$

or

$$\sum_{n\in A_i} \operatorname{Im} \hat{f}(n)\hat{g}(n) = + \text{ or } -\infty \text{ if } i = 3 \text{ or } 4 .$$

So S may be taken to be A_i for some i .)

Now define A by

$$A = \{h \in L^p : \text{sp } h \subset S\}.$$

Theorem 2.2 implies that A is not invariantly complemented in L^p, hence A is uncomplemented in L^p by Theorem 2.1. Then $A^\perp$ is an invariant uncomplemented subspace of L^q, so the case $2 < p < \infty$ is thus taken care of also.

Q.E.D.

We shall now give some characterizations of Sidon sets. Toward this end, we need the following

Lemma 2.4. Let $\varphi \in C(G)$, and suppose that for every subset S of Γ, $\sum_{\tau \in S} \hat{\varphi}(\tau)(\tau, \cdot)$ is the Fourier series of a member of $C(G)$.

Then $\sum_{\tau \in \Gamma} |\hat{\varphi}(\tau)| < \infty$.

(This lemma follows easily from well known facts; for the sake of completeness, we include the proof.)

Proof. First we note the following: if $h \in L^\infty(G)$ and $\hat{h}(\tau) \geq 0$, all $\tau \in \Gamma$, then $\sum_{\tau \in \Gamma} \hat{h}(\tau) < \infty$. For, let S be a finite subset of Γ. It follows from 2.6.8, p. 53, of [15] that we may choose a trigonometric polynomial p, $\hat{p}(\tau) \geq 0$, all $\tau \in \Gamma$, $\hat{p}(s) = 1$, all $s \in S$, and $\|p\|_{L^1(G)} \leq 2$. We then have

$$\begin{aligned} 2\|h\|_\infty &\geq \left|\int_G p(-x)h(x)dx\right| \\ &= \sum_{\tau \in \Gamma} \hat{p}(\tau)\hat{h}(\tau) \\ &\geq \sum_{\tau \in S} \hat{p}(\tau)\hat{h}(\tau) = \sum_{\tau \in S} \hat{h}(\tau) . \end{aligned}$$

Since S was an arbitrary finite subset of Γ, we thus have

$$\sum_{\tau\in\Gamma} \hat{h}(\tau) \leq 2\|h\|_\infty < \infty \ .$$

Now we assume that $\varphi \in C(G)$, and that $\sum_{\tau\in S} \hat{\varphi}(\tau)(\tau, \cdot)$ is the Fourier series of a member of $C(G)$, each $S \subset \Gamma$. Suppose that $\sum_{\tau\in\Gamma} |\hat{\varphi}(\tau)| = \infty$. Then either

$$\sum_{\tau\in\Gamma} |\text{Re}\, \hat{\varphi}(\tau)| = \infty$$

or

$$\sum_{\tau\in\Gamma} |\text{Im}\, \hat{\varphi}(\tau)| = \infty \ .$$

Suppose the former; let

$$S = \{\tau \in \Gamma : \text{Re}\, \hat{\varphi}(\tau) \geq 0\}.$$

Then either

$$\sum_{\tau\in S} \text{Re}\, \hat{\varphi}(\tau) = \infty ,$$

or

$$\sum_{\tau\in\Gamma\sim S} \text{Re}\, \hat{\varphi}(\tau) = -\infty .$$

Again supposing the first possibility, our hypotheses imply that $\sum_{\tau\in S} \hat{\varphi}(t)(\tau, .)$ is the Fourier series of a continuous function, f.

Then $\sum_{\tau\in S} \text{Re}\, \hat{\varphi}(\tau)(\tau,.)$ is the Fourier series of

$$h(x) = \frac{f(x) + \overline{f(-x)}}{2} ,$$

a continuous function with non-negative Fourier coefficients. By the remark made at the beginning of this proof, we must therefore have

$$\sum_{\tau\in\Gamma} |\hat{h}(\tau)| < \infty \ .$$

Thus

$$\sum_{\tau \in \Gamma} \hat{h}(\tau) = \sum_{\tau \in S} \hat{h}(\tau) = \sum_{\tau \in S} \mathrm{Re}\ \hat{\varphi}(\tau) < \infty ,$$

a contradiction. The other possibilities are handled similarly.

Q.E.D.

Corollaries of Theorem 2.1 and Lemma 2.4:

Corollary 2.5. *If A is an invariant subspace of* $C(G)$ *, and every invariant subspace of A is complemented in A , then* sp A *is a Sidon set*.

Proof. Let $S \subset \Gamma$. Then

$$D = \{f \in C(G) : \mathrm{sp}\ f \subset S \cap \mathrm{sp}\ A\}$$

is complemented in A . Hence by Theorem 2.1 and the Spectral Synthesis Corollary, if $\varphi \in A$,

$$\sum_{\tau \in S} \hat{\varphi}(\tau)(\tau, .) = \sum_{\tau \in S \cap \mathrm{sp}A} \hat{\varphi}(\tau)(\tau, .)$$

is the Fourier series of a member of $C(G)$. Since S was arbitrary, Lemma 2.4 implies that $\sum_{\tau \in \Gamma} |\hat{\varphi}(\tau)| < \infty$.

Corollary 2.6. *If A is an invariant subspace of* $L^1(H)$, H *an arbitrary locally compact abelian group, and if A is of infinite codimension (i.e.,* hA *infinite), then* $L^1(H)/A$ *is not isomorphic to Hilbert space, since its dual contains a copy of* ℓ .

Proof.* Let B = almost-periodic continuous functions belonging to $A^{\perp}$. If L^1/A were isomorphic to Hilbert space, $A^{\perp}$ = dual of L^1/A would be also; hence B, a closed subspace of $A^{\perp}$, would be isomorphic to Hilbert space. So every invariant subspace of B would be complemented in B. But B is naturally isomorphic to an invariant subspace of $C(\breve{H})$, $\breve{H}$ = the Bohr compactification of H (see the paragraph preceding Lemma 1.2). Thus we obtain that sp B, regarded as a subset of the dual of $\breve{H}$, is a Sidon set, by applying Corollary 2.5 to the compact group $\breve{H}$.

But sp B = hA is infinite. Hence ℓ (spB) is non-reflexive, and B is isomorphic to ℓ (spB), as was mentioned in the introductory remarks on Sidon sets.

So B could not be isomorphic to Hilbert space.

Theorem 2.7. Let A be an invariant weak* closed subspace of $L^{\infty}(G)$. Then the following are equivalent:

(a) Every weak* closed invariant subspace of A is (norm)

---

* Added in proof: The final clause of the statement of Corollary 2.6, added at the referee's suggestion, renders our proof completely superfluous. However, an argument similar to that given in the proof has yielded the following result: Suppose that hA is compact, and that for every closed set $E \subset hA$ there is a projection from $L^1(H)/A$ onto $\{f \in L^1(H)/A: \hat{f}=0 \text{ on } E\}$. Then hA is a Helson set; i.e. $C(hA) = \{\hat{f}: f \in L^1(H)/A\}$. (See Rosenthal, H.P.: "Caracterisation d'ensembles de Helson, par l'existence de certains projecteurs". C.R. Acad. Sci. Paris, t. 262 (31 Jan. 66) pp. 286-288).

<u>complemented in</u> A .

(b) sp A <u>is a Sidon set</u>.

(c) $\sum_{\tau \in \Gamma} |\hat{f}(\tau)\hat{g}(\tau)| < \infty$, all $f \in A$, $g \in L^1(G)$.

<u>Proof</u>. (a) $\Longrightarrow$ (b). Let D be an arbitrary invariant subspace of $A \cap C(G)$. Let

$$B = \{h \in L^{\infty}(G) : \text{sp } h \subset \text{sp } D\} .$$

By the Spectral Synthesis Theorem, $D = B \cap C(G)$. Now by Theorem 1.1, since B is a weak* closed invariant subspace of A , and B is norm-complemented in A , B is invariantly complemented.

If we let T be the invariant projection, then by Lemma 1.2 there is a finite measure μ on G , such that $Tf = f * \mu$, all continuous $f \in A$. $T|(A \cap C(G))$ is thus a projection from $A \cap C(G)$ onto $B \cap C(G)$. Hence $A \cap C(G)$ satisfies the hypothesis of Corollary 2.5 so $\text{sp}(A \cap C(G)) =$ sp A is a Sidon set.

(b) $\Longrightarrow$ (a) and (b) $\Longrightarrow$ (c) are immediate, using the equivalent definitions of a Sidon set mentioned in the introduction to this section.

(c) $\Longrightarrow$ (b). We are assuming that

$$\sum_{\tau \in \Gamma} |\hat{f}(\tau)\hat{g}(\tau) < \infty \text{ , all } g \in L^1(G) \text{ , } f \in A \text{ .}$$

Now fix $f \in A$. Then the τ's such that $\hat{f}(\tau) \neq 0$ form a denumerable set, $\{\tau_n\}_{n=1}^{\infty}$.

Define

$$T_n : L^1(G) \to A(G)$$

by

$$T_n(g) = \sum_{k=1}^{n} \hat{f}(\tau_k)\hat{g}(\tau_k)\,(\tau_k, .) .$$

The T_n 's thus form a pointwise convergent family of bounded linear operators from $L^1(G)$ into the absolutely convergent Fourier series $A(G)$, so by the uniform boundedness principle there is a constant K such that

$$\|T_n\| \leq K , \text{ all } n .$$

so

$$\sum_{k=1}^{n} |\hat{f}(\tau_k)\hat{g}(\tau_k)| \leq K\|g\|_{L^1(G)} , \text{ all } n , g \in L^1(G) .$$

For fixed n , choose

$$g \in L^1(G) , \|g\|_{L^1(G)} \leq 2 , \hat{g}(\tau_k) = 1 , 1 \leq k \leq n .$$

(Such a g exists by Theorem 2.6.8, p. 53, of [15] .) Then

$$\sum_{k=1}^{n} |\hat{f}(\tau_k)| \leq 2K .$$

This holds for all integers n , whence

$$\sum_{k=1}^{\infty} |\hat{f}(\tau_k)| \leq 2K < \infty .$$

Q.E.D.

The final theorem of this section is the characterization of invariant Hilbert subspaces of $L^p(G)$ mentioned in the introduction. To handle the case $1 \leq p < 2$, the following lemma is needed. (This also completes the proof of Corollary 2.3.)

Lemma 2.8. Consider $\{w_\tau\}_{\tau \in \Gamma}$, where $w_\tau \geq 0$ for all τ . Suppose $\sum_{\tau \in \Gamma} w_\tau |\hat{f}(\tau)| < \infty$ for all $f \in C(G)$. Then $\sum_{\tau \in \Gamma} w_\tau^2 < \infty$.

Proof. It suffices to show that $\sum_{\tau\in S} w_{\tau}^{2} < \infty$, for each countable $S \subset \Gamma$. So let S be countable, $\tau_0, \tau_1, \tau_2, \ldots$ the members of S.

Define $T_n : C(G) \to A(G)$ by

$$(T_n f)(x) = \sum_{j=0}^{n} w_{\tau_j} \hat{f}(\tau_j)(\tau_j, x) .$$

($A(G)$ = absolutely convergent Fourier series on G .)

$\{T_n\}$ is a pointwise convergent sequence of bounded linear operators, hence by the uniform boundedness principle there is a constant K,

$$\sum_{j=1}^{n} w_{\tau_j} |\hat{f}(\tau_j)| \leq K\|f\|_{\infty} \text{, all integers } n \text{, all } f \in C(G) .$$

Now fix n an integer; let $r_0, \ldots, r_n$ be the first n+1 Rademacher functions. (For the definition and basic property of these functions, see the introduction to this section.) For each $t \in [0, 1]$, define

$$P_t(x) = \sum_{j=0}^{n} r_j(t) w_{\tau_j} (\tau_j, x) \text{, all } x \in G .$$

P_t is thus a trigonometric polynomial on G ; if $f \in C(G)$, then

$$\begin{aligned}
\left|\int_G P_t(x) f(-x)\,dx\right| &= \left|\sum_{j=0}^{n} r_j(t) w_{\tau_j} \hat{f}(\tau_j)\right| \\
&\leq \sum_{j=0}^{n} |r_j(t) w_{\tau_j} \hat{f}(\tau_j)| \\
&= \sum_{j=0}^{n} w_{\tau_j} \hat{f}(\tau_j) \qquad (|r_j(t)| = 1, \text{ all } t, j) \\
&\leq K\|f\|_{\infty} .
\end{aligned}$$

K is independent of t and of n. So for each $t \in [0,1]$,

$\|p_t\|_{L^1(G)} \leq K$. Thus

$$\int_G \left(\int_0^1 |p_t(x)|dt\right) dx = \int_0^1 \left(\int_G |p_t(x)|\,dx\right) dt \text{ (by Fubini's theorem)}$$

$$\leq \int_0^1 K\,dt = K .$$

Define $h(x)$ on G by

$$h(x) = \int_0^1 |p_t(x)|\,dt .$$

Then $h(x) \geq 0$ for all $x \in G$, and $\int_G h(x)dx \leq K$. Hence for some $x_0 \in G$, $h(x_0) \leq K$. Now as a function of t , $p_t(x_0)$ is just a Rademacher polynomial. Thus by Theorem 2.0,

$$\left(\int_0^1 |p_t(x_0)|^2 dt\right)^{\frac{1}{2}} \leq \frac{1}{A_1}\int_0^1 |p_t(x_0)| \leq \frac{B_1}{A_1} K ,$$

where A_1 and B_1 are defined in Theorem 2.0.

A_1 is, of course, independent of n , and

$$\left(\int |p_t(x_0)|^2 dt\right)^{\frac{1}{2}} = \left(\sum_{j=0}^{n} |w_{\tau_j}(\tau_j, x_0)|^2\right)^{\frac{1}{2}}$$

$$= \left(\sum_{j=0}^{n} w_{\tau_j}^{\,2}\right)^{\frac{1}{2}} \quad \text{(since } |(\tau, x_0)| = 1 \text{ for all } \tau \in \Gamma) .$$

Hence

$$\sum_{j=0}^{n} w_{\tau_j}^{\,2} \leq \left(\frac{B_1}{A_1} K\right)^2 ,$$

so

$$\sum_{j=0}^{\infty} w_{\tau_j}^{\;2} \leq \left(\frac{B_1}{A_1} K \right)^2 < \infty .$$

Q.E.D.

Remark. In the case in which Γ can be ordered, this lemma is an immediate consequence of the theorem on p. 222 of [15].

Theorem 2.9. Let A be an invariant subspace of $L^p(G)$, $1 \leq p < \infty$, $p \neq 2$. Then the following are equivalent:

(a) A is isomorphic to Hilbert space.

(b) Every invariant subspace of A is complemented in A.

(c) $\sum_{\tau \in \Gamma} |\hat{f}(\tau)\hat{g}(\tau)| < \infty$, all $f \in A$, $g \in L^q(G)$.

(d) $A \subset L^2(G)$, and the L^r norms on A are all equivalent, $1 \leq r \leq \max\{p,2\}$.

Proof. (a) $\Longrightarrow$ (b). Immediate.

(b) $\Longrightarrow$ (c). Let $S \subset \Gamma$. Then

$$D = \{h \in L^p(G) : \text{sp } h \subset S \cap \text{sp } A\}$$

is an invariant subspace of A. Hency, by the Spectral Synthesis Corollary and Theorem 2.1, if $f \in A$, then

$$\sum_{\tau \in S} \hat{f}(\tau)(\tau, \cdot) = \sum_{\tau \in S \cap \text{sp}A} \hat{f}(\tau)(\tau, .)$$

is the Fourier series of h, a member of $L^p(G)$. Since $h \in L^p$, if we take $g \in L^q$, then $h*g \in C(G)$.

Thus

$$\sum_{\tau \in \Gamma} \hat{h}(\tau)\hat{g}(\tau)(\tau, .) = \sum_{\tau \in S} \hat{f}(\tau)\hat{g}(\tau)(\tau, .)$$

is the Fourier series of a continuous function, for each $f \in A$, $g \in L^q$, $S \subset \Gamma$. So fix $f \in A$, $g \in L^q$; then Lemma 2.4 tells us that $\sum_{\tau \in \Gamma} |\hat{f}(\tau)\hat{g}(\tau)| < \infty$.

(c) $\Longrightarrow$ (d). We must do $1 \le p < 2$, $2 < p < \infty$ separately.

Case (i). $1 \le p < 2$.

Fix $f \in A$. Since $C(G) \subset L^q(G)$, we have that $\sum_{\tau \in \Gamma} |\hat{f}(\tau)|^2 < \infty$, by Lemma 2.8 (letting $w_\tau = |\hat{f}(\tau)|$).

Thus $f \in L^2(G)$, so $A \subset L^2(G)$. Finally, A1 and A2 of the Appendix imply, for $1 \le p < 2$, that any closed subspace of $L^p(X)$ that is contained in $L^2(X)$ must have all L^r norms equivalent on it, $1 \le r \le 2$, X a finite measure space. Hence the L^r norms are all equivalent on A.

Case (ii). $2 < p < \infty$.

Define $T : L^q \times A \to A(G)$ by

$$T(g,f)(x) = g*f(x) = \sum_{\tau \in \Gamma} \hat{g}(\tau)\hat{f}(\tau)(x,\tau).$$

T is a bilinear map. Now fix $f \in A$; then $g \to T(g,f)$ is a bounded linear transformation from $L^q(G)$ to $A(G)$, by the closed graph theorem (using uniqueness of Fourier transform). Similarly, fix $g \in L^q(G)$; then $f \to T(g,f)$ is a bounded linear map from A to $A(G)$ by the closed graph theorem. An easy application of the uniform boundedness principle now yields that there is a constant K, so that

$$\sum_{\tau \in \Gamma} |\hat{g}(\tau)\hat{f}(\tau)| = \|T(g,f)\|_{A(G)} \le K\|f\|_p\|g\|_q, \text{ all } f \in A,\ g \in L^q(G).$$

Now fix $f \in A$, f a trigonometric polynomial; let $\tau_0, \tau_1, \ldots, \tau_k$

be the spectrum of f . Fix $t_0 \in [0,1]$ and let $r_0, r_1, \ldots, r_k$ be the first $k+1$ Rademacher functions.

(*)

$$\|f\|_{L^p(G)}$$

$$= \sup_{\substack{g \in L^q(G) \\ \|g\|_q = 1}} \left| \int_G f(x)g(-x)dx \right|$$

$$= \sup_{\substack{g \in L^q(G) \\ \|g\|_q = 1}} \left| \sum_{n=0}^{k} \hat{f}(\tau_n)\hat{g}(\tau_n) \right|$$

$$\leq \sup_{\substack{g \in L^q(G) \\ \|g\|_q = 1}} \sum_{n=0}^{k} |\hat{f}(\tau_n)\hat{g}(\tau_n)|$$

$$= \sup_{\substack{g \in L^q(G) \\ \|g\|_q = 1}} \sum_{n=0}^{k} |r_n(t_0)\hat{f}(\tau_n)\hat{g}(\tau_n)|$$

$$\leq K \left\| \sum_{n=0}^{k} r_n(t_0)\hat{f}(\tau_n)(\tau_n, \cdot) \right\|_{L^p(G)} .$$

The last inequality follows from the fact that

$$\|T(g,h)\|_{A(G)} \leq K\|h\|_p \|g\|_q ,$$

where

$$h(x) = \sum_{n=0}^{k} r_n(t_0)\hat{f}(\tau_n)(\tau_n, x) .$$

So far t_0 has been arbitrary. We shall now use Theorem 2.0 to select t_0 appropriately. Let

$$g(x,t) = \sum_{n=0}^{k} \hat{f}(\tau_n)(\tau_n, x)r_n(t)$$

defined for $x \in G$, $t \in [0,1]$. If we fix x, then $t \to g(x,t)$ is just a Rademacher polynomial. Thus by Theorem 2.0,

$$\left(\int_0^1 |g(x,t)|^p dt\right)^{\frac{1}{p}}$$

$$\leq B_p \left(\int_0^1 |g(x,t)|^2 dt\right)^{\frac{1}{2}}$$

$$= B_p \left(\sum_{n=0}^{k} |(\gamma_n, x) f(\gamma_n)|^2\right)^{\frac{1}{2}}$$

$$= B_p \left(\sum_{n=0}^{k} |\hat{f}(\gamma_n)|^2\right)^{\frac{1}{2}} = B_p \|f\|_{L^2(G)} .$$

B_p is the constant defined in Theorem 2.0, and is hence independent of x and f. By Fubini's theorem,

$$\int_0^1 \left(\int_G |g(x,t)|^p dx\right) dt$$

$$= \int_G \left(\int_0^1 |g(x,t)|^p dt\right) dx$$

$$\leq \int_G \left(B_p{}^p \|f\|^p{}_{L^2(G)}\right) dx = B_p{}^p \|f\|^p{}_{L^2(G)} .$$

Let

$$q(t) = \int_G |g(x,t)|^p dx .$$

We have $q(t) \geq 0$,

$$\int_0^1 q(t) dt \leq B_p{}^p \|f\|^p{}_{L^2(G)} .$$

Then for some t_0 we have

$$q(t_0) \leq B_p{}^p \|f\|^p{}_{L^2(G)}$$

So

$$(q(t_0))^{\frac{1}{p}}$$

$$= \left\| \sum_{n=0}^{k} r_n (t_0) \hat{f}(\tau_n)(\tau_n, .) \right\|_{L^p(G)}$$

$$\leq B_p \|f\|_{L^2(G)}$$

Thus by (✱),

$$\|f\|_{L^p(G)} \leq K \left\| \sum_{n=0}^{k} r_n(t_0) \hat{f}(\tau_n)(\tau_n, .) \right\|_{L^p(G)}$$

$$\leq K B_p \|f\|_{L^2(G)}$$

for all trigonometric polynomials $f \in A$ (K , B_p were independent of f , of course). By the remark following the Spectral Synthesis Theorem the trigonometric polynomials in A are dense in A . Hence, for every $f \in A$,

$$\|f\|_{L^p(G)} \leq K B_p \|f\|_{L^2(G)} .$$

Finally, by A2 of the Appendix, the L^r norms on A are all equivalent, $1 \leq r \leq p$.

(d) $\Longrightarrow$ (a). If (d) is true, A is isomorphic to a subspace of $L^2(G)$, hence A is isomorphic to Hilbert space.

Remark. In Theorem 3 of [12] it is proved that (a) and (d) are equivalent for $p > 2$, an arbitrary subspace of $L^p(X, \mathcal{S}, \mu)$, $(X, \mathcal{S}, \mu)$ an arbitrary finite measure space. The remaining equivalences of the theorem are new, in so far as we know.

Corollary 2.10. Let $2 < p < \infty$, A an invariant subspace of $L^p(G)$, A isomorphic to Hilbert space. Then A is complemented in $L^p(G)$.

Proof. This follows immediately from (d) of Theorem 2.9 and A3 of the Appendix.

Our next result characterizes those Hilbert subspaces of $L^p(G)$ that are complemented.

We remark that (b) and (c) of 2.11 are more stringent that (b) and (c) of 2.9 for some values of $p<2$, since there are indeed Hilbert subspaces of L^p that are uncomplemented (see Remark 2).

Theorem 2.11. *Let* A *be an invariant subspace of* $L^\alpha(G)$, $1 < \alpha < \infty$, $\alpha \neq 2$, $\frac{1}{\alpha} + \frac{1}{\beta} = 1$. *Then the following are equivalent*:

(a) A *is isomorphic to Hilbert space and complemented in* $L^\alpha(G)$.

(b) *Every invariant subspace of* A *is complemented in* $L^\alpha(G)$.

(c) $\sum_{\tau \in \operatorname{sp} A} |\hat{f}(\tau)\hat{g}(\tau)| < \infty$, *all* $f \in L^\alpha(G)$, $g \in L^\beta(G)$.

(d) $A \subset L^s$, $s = \max\{\alpha, \beta\}$, and the L^r norms on A are all *equivalent*, $1 \leq r \leq s$.

Proof. (a) $\Longrightarrow$ (b). Immediate.

(b) $\Longrightarrow$ (c): By Theorem 2.1, every invariant subspace of A is invariantly complemented in $L^\alpha(G)$, and hence in A, by the very definition of "complemented". Thus (b) of Theorem 2.9 holds, so by (c) of Theorem 2.9,

$$\sum_{\tau \in \Gamma} |\hat{f}(\tau)\hat{g}(\tau)| < \infty, \text{ all } f \in A,\ g \in L^\beta(G).$$

By hypothesis, A is invariantly complemented in $L^\alpha(G)$; by the Spectral Synthesis Corollary the invariant projection P from $L^\alpha(G)$ onto A has the property that the Fourier series of Pf is $\sum_{\tau \in \operatorname{sp} A} \hat{f}(\tau)(\tau, \cdot)$, all $f \in L^{\alpha}(G)$. Thus

$$\sum_{\tau \in \Gamma} |\widehat{Pf}(\tau)\hat{g}(\tau)| = \sum_{\tau \in spA} |\hat{f}(\tau)\hat{g}(\tau)| < \infty, \text{ all } f \in L^{\alpha}(G), g \in L^{\beta}(G).$$

(c) $\Longrightarrow$ (d) . Suppose $\beta = \max\{\alpha, \beta\}$. Define

$$A_1 = \{g \in L^{\beta}(G) \ : \ sp\ g \subset sp\ A\} .$$

Then Condition (c) of Theorem 2.9 holds, where the "A" for Theorem 2.9 is A_1 , the "p" of Theorem 2.9 is β . So by (d) of Theorem 2.9 the L^r norms on A_1 are all equivalent, $1 \leq r \leq \beta$. It thus follows that A_1 is a closed subspace of $L(G)$. Hence by the Spectral Synthesis Theorem, $A_1 = A$. We have already established that the L^r norms on A_1 are all equivalent, so we are finished.

If $\alpha = \max\{\alpha, \beta\}$, then (c) of Theorem 2.9 holds, where the "A" of Theorem 2.0 is A , and the "p" of Theorem 2.9 is α . Thus in this case (d) follows immediately from (d) of Theorem 2.9.

(d) $\Longrightarrow$ (a). Since the L^2 norm on A is equivalent to the L^{α} norm by hypothesis, A is isomorphic to Hilbert space. Moreover, (d) implies that

$$A = \{f \in L^s(G) : sp\ f \subset sp\ A\}.$$

Thus by Corollary 2.10, A as a subspace of $L^s(G)$ is complemented in $L^s(G)$. If $s = \alpha$, we are finished. Otherwise, Theorem 2.1 says that A is invariantly complemented in $L^s(G) = L^{\beta}(G)$. This means that

$$L^{\beta}(G) = A \oplus B ,$$

where

$$B = \{g \in L^{\beta}(G) : sp\ g \cap sp\ A = \emptyset\}.$$

Then

$$L^{\infty}(G) = A^{\perp} \oplus B^{\perp} .$$

But $B^{\perp} = A$, where A is regarded as a subspace of $L^{\infty}(G)$. (This follows by the Spectral Synthesis Theorem, since sp $B^{\perp}$ = sp A .) Hence A is complemented in $L^{\infty}(G)$.

<u>Remark</u> 1. (d) of Theorem 2.11 shows that a fifth condition equivalent to (a), (b), (c), and (d) of Theorem 2.11 is that sp A is a p-Sidon set, as defined in [6] .

<u>Remark</u> 2. In [17] certain subsets of the integers are defined as follows: <u>E is said to be of type</u> $\Lambda(s)$, <u>or</u> $E \in \Lambda(s)$, <u>if the</u> L^r <u>norms are equivalent on all trigonometric polynomials</u> f <u>with</u> sp $f \subset E$, <u>for</u> $1 \leq r \leq s$.

It is proved in [17] that for every $\epsilon > 0$ there is a set E such that

$$E \in \Lambda(4) , E \notin \Lambda(4+\epsilon) .$$

Now fix such an E , and let $A = \{f \in L^4(\mathbb{T}) : \text{sp } f \subset E\}$. Then our Corollary 2.11 implies that A is not complemented in $L^{\delta}(\mathbb{T})$, where $\frac{1}{\delta} + \frac{1}{4+\epsilon} = 1$. For A is isomorphic to Hilbert space; hence if A were complemented in $L^{\delta}(\mathbb{T})$, $E \in \Lambda(4+\epsilon)$ by (d) of Corollary 2.11, where "α" of Corollary 2.11 is δ, "β" of Corollary 2.11 is $4+\epsilon$. Thus we see that Corollary 2.10 fails for $1 \leq p < 2$. Indeed, we have that for every $1 \leq p < \frac{4}{3}$, there is an invariant subspace of $L^p(\mathbb{T})$, isomorphic to Hilbert space and uncomplemented in $L^p(\mathbb{T})$. (This thus answers in the negative a question raised in [12] .)

<u>Remark</u> 3. Theorems 2.7 and 2.9 and Corollary 2.5 yield the following,

in terms of the "Λ" sets defined in [17]:

Let $A \subset \mathcal{B}$, ($\mathcal{B}$ as defined in the introduction to this section), A invariant, and let every invariant subspace of A be complemented in A. Or let $A \subset L^{\infty}$, A weak* closed invariant, and let every invariant weak* closed subspace of A be complemented in A.

(a) If $A \subset L^p$, $1 \leq p < 2$, then $\text{sp}\, A \in \Lambda(2)$.

(b) If $A \subset L^p$, $2 < p < \infty$, then $\text{sp}\, A \in \Lambda(p)$.

(c) If $A \subset C(G)$ or if $A \subset L^{\infty}(G)$, then $\text{sp}\, A$ is a Sidon set.

Hence $\text{sp}\, A \in \Lambda(r)$, all $r < \infty$, by A2 of the Appendix and p. 128 of [15].

The final two results of this section show that our results dealing with invariant Hilbert subspaces are quite properly stated for compact groups; they have no content for non-compact groups.

(We are indebted to Professor Katznelson for suggesting the proofs of the following two results; our original argument was much longer, and turned on the results of section III).

<u>Theorem</u> 2.12. <u>Let</u> H <u>be a non-compact locally compact abelian group, and let</u> A <u>be a non-zero invariant subspace of</u> $L^p(H)$, $1 \leq p < \infty$, $p \neq 2$. <u>Then</u> A <u>has a subspace isomorphic to</u> ℓ^p.

Proof: (note that $\|\cdot\|_p = \|\cdot\|_{L^p(H)}$)

Let $f \in A$, $\|f\|_p = 1$. f is now fixed throughout. Choose, for each integer $n \geq 1$, a compact set C_n so that

$$\|\chi_{C_n} \cdot f\|_p \geq 1 - \frac{1}{2^{n+1}} .$$

Now choose (by induction) a sequence $h_1, h_2, \ldots$ in H, such that

$$(h_i + C_i) \cap (h_j + C_j) = \emptyset \text{ for } i \neq j .$$

(We can do this, since H is assumed to be non-compact.) Define

$$f_n(x) = f(x - h_n), \quad \text{for all} \quad x \in H$$

($f_n \in A$ for all n, of course.)

Now we claim that the closed linear span of the f_n's is isomorphic to ℓ^p. (The following part of the argument is similar to an argument given in [12].)

It suffices to show that there exist positive constants k and K so that for any n complex numbers $a_1, \ldots, a_n$,

$$k\left(\sum_{i=1}^{n} |a_i|^p\right)^{\frac{1}{p}} \leq \left\|\sum_{i=1}^{n} a_i f_i\right\|_p \leq K\left(\sum_{i=1}^{n} |a_i|^p\right)^{\frac{1}{p}} .$$

First, define

$$\varphi_n = \chi_{h_n + C_n} \cdot f_n .$$

Thus $\frac{3}{4} \leq \|\varphi_n\|_p \leq 1$, the φ_n's are disjointly supported, and

$$\|\varphi_n - f_n\|_p = \|\chi_{C_n} \cdot f - f\|_p \quad \text{(by invariance of Haar measure)}$$

$$\leq \frac{1}{2^{n+1}} .$$

Now let $a_1, \ldots, a_n$ be n complex numbers. Then

$$\left\|\sum_{i=1}^{n} a_i (\varphi_i - f_i)\right\|_p$$

$$\leq \sum_{i=1}^{n} |a_i| \, \|\varphi_i - f_i\|_p \quad \text{(by Minkowski's inequality)}$$

$$\leq \left(\sum_{i=1}^{n} |a_i|^p\right)^{\frac{1}{p}} \left(\sum_{i=1}^{n} \|\varphi_i - f_i\|_p^{\,q}\right)^{\frac{1}{q}}$$

(by Hölder's inequality)

$$\leq \left(\sum_{i=1}^{n} |a_i|^p\right)^{\frac{1}{p}} \sum_{i=1}^{n} \|\varphi_i - f_i\|_p$$

$$\leq \left(\sum_{i=1}^{n} |a_i|^p\right)^{\frac{1}{p}} \sum_{i=1}^{n} \frac{1}{2^{i+1}} \leq \frac{1}{2} \left(\sum_{i=1}^{n} |a_i|^p\right)^{\frac{1}{p}} .$$

Thus

$$\left\|\sum a_i f_i\right\|_p$$

$$\leq \left\|\sum a_i (f_i - \varphi_i)\right\|_p + \left\|\sum a_i \varphi_i\right\|_p \quad \text{(by Minkowski's inequality)}$$

$$\leq \frac{1}{2}\left(\sum |a_i|^p\right)^{\frac{1}{p}} + \left\|\sum a_i \varphi_i\right\|_p$$

$$= \frac{1}{2}\left(\sum |a_i|^p\right)^{\frac{1}{p}} + \left(\sum |a_i|^p \|\varphi_i\|_p^{\,p}\right)^{\frac{1}{p}}$$

(because the φ_i have disjoint supports)

$$\leq \frac{1}{2}\left(\sum |a_i|^p\right)^{\frac{1}{p}} + \left(\sum |a_i|^p\right)^{\frac{1}{p}}$$

$$\leq \frac{3}{2}\left(\sum |a_i|^p\right)^{\frac{1}{p}} .$$

Finally,

$$\left\|\sum a_i f_i\right\|_p \geq \left\|\sum a_i \varphi_i\right\|_p - \left\|\sum a_i (f_i - \varphi_i)\right\|_p$$

$$\geq \left(\sum |a_i|^p \|\varphi_i\|_p^{\,p}\right)^{\frac{1}{p}} - \frac{1}{2}\left(\sum |a_i|^p\right)^{\frac{1}{p}}$$

$$\geq \frac{3}{4}\left(\sum |a_i|^p\right)^{\frac{1}{p}} - \frac{1}{2}\left(\sum |a_i|^p\right)^{\frac{1}{p}}$$

$$= \frac{1}{4}\left(\sum |a_i|^p\right)^{\frac{1}{p}} .$$

Thus we have demonstrated that

$$\frac{1}{4}\left(\sum_{i=1}^{n} |a_i|^p\right)^{\frac{1}{p}} \leq \left\|\sum_{i=1}^{n} a_i f_i\right\|_p \leq \frac{3}{2}\left(\sum_{i=1}^{n} |a_i|^p\right)^{\frac{1}{p}} ,$$

so the proof of Theorem 2.12 is complete.

Corollary 2.13. Let A and H as in theorem 2.12. Then A is not isomorphic to Hilbert space; moreover, when p=1, A is not reflexive.

Proof: By Theorem 7, p. 205, of [1], ℓ^p is not isomorphic to ℓ^2 for $p \neq 2$. Hence A has a subspace non-isomorphic to Hilbert space, so A itself is not isomorpnic to Hilbert space. Finally, ℓ^1 is not reflexive; thus for p= 1, A is not reflexive, since A has a non-reflexive subspace, namely one that is isomorphic to ℓ^1.

Q.E.D.

SECTION III

PROJECTIONS ONTO TRANSLATION - INVARIANT SUBSPACES OF $L^p(\mathbb{R}^n)$

This section deals with complementation in invariant subspaces of $L^p(G)$, where G is a locally compact abelian group, and $1<p<2$. Our main result is Theorem 3.8, where we show that if A is a non-zero invariant subspace of $L^p(\mathbb{R}^n)$, then there is an invariant subspace of A that is uncomplemented in A. (Hence A is not isomorphic to Hilbert space; see Theorem 2.12 for a different proof of this fact.)

The first step in achieving Theorem 3.8 is Lemma 3.1, where we show that if A and B are invariant subspaces of $L^p(G)$, with $A \subset B$ and A complemented in B with norm $\leq C$, then A is invariantly complemented in B with norm $\leq C$. (This result holds for non-abelian amenable groups also; see the remarks following Lemma 3.1. This also holds for all p, $1<p<\infty$.)

Next we show in Theorem 3.2 that is $u \in L^p(\mathbb{R}^n)$, $u \neq 0$, $1<p<2$, then there exists an open set S such that χ_S does not multiply $\hat{u}$. I.e., for no $g \in L^p(\mathbb{R}^n)$ is it true that $\chi_S \hat{f} = \hat{g}$ a. e.

The proof of Theorem 3.8 is then completed by means of Lemmas 3.6 and 3.7. Here we show, for an arbitrary group G and $1<p<2$, that if $u \in L^p(G)$ and S is an open subset of Γ such that χ_S does not multiply $\hat{u}$, then if A is the smallest invariant subspace of L^p containing u, there is an invariant subspace B of A that is uncomplemented in A, namely

$$B = \{f \in A : \hat{f}(x) = 0 \text{ for almost all } x \notin S\}.$$

As side results we obtain in Corollary 3.4 that if $u \in L^1 \cap L^2$, there exists an open bounded set S such that $\chi_S \hat{u}$ is not the transform of a function in $L^p(\mathbb{R}^n)$ for any $1<p<2$. We show in Corollary 3.5 that there exists an open bounded set S such that

$$A_p = \{f \in L^p(\mathbb{R}^n) : \hat{f}(x) = 0 \text{ for almost all } x \notin S\}$$

is uncomplemented in L^p, all $1<p<2$.

Before we proceed with the work of this section we need a few definitions and some standard facts.

If A is an invariant subspace of $L^p(\mathbb{R}^n)$, $1<p<2$, and E is a measurable subset of $\mathbb{R}^n$, we say that χ_E <u>multiplies</u> $\hat{A}$ <u>with norm</u> $\leq C$ if there exists a bounded linear operator $T: A \to A$ such that $\widehat{Tf} = \chi_E \hat{f}$ a. e., all $f \in A$, and $\|T\| \leq C$.

If $u \in L^p(\mathbb{R}^n)$ we say that χ_E <u>multiplies</u> $\hat{u}$ <u>with norm</u> $\leq C$, if there exists a $g \in L^p$ with $\|g\|_p \leq C$ and $\hat{g} = \chi_E \hat{u}$ a. e. (As usual, "a.e." stands for "almost everywhere"; we will also use the notation "a.a." to stand for "almost all".)

Finally, we need the following two well known facts.

<u>Fact</u> 1. <u>Let</u> G <u>be given,</u> A <u>an invariant subspace of</u> $L^p(G)$, $1 \leq p<\infty$; <u>let</u> μ <u>be a finite measure on</u> G. <u>Then</u> $\mu * f \in A$ <u>for each</u> $f \in A$.

<u>Fact</u> 2. <u>If</u> $f \in L^p(G)$, $1 \leq p \leq 2$, <u>then</u> $\hat{f}$ <u>is a well defined member of</u> $L^q(\Gamma)$, <u>with</u>

$$\|\hat{f}\|_{L^q(\Gamma)} \leq \|f\|_{L^p(G)} .$$

(See p. 152 of [8].)

We begin the work of Section III with Lemma 3.1, which shows that the existence of a projection onto an invariant subspace of L^p, $1<p<\infty$, implies the existence of an invariant projection onto that subspace.

Lemma 3.1. *Let* A, B *be invariant subspaces of* $L^p(G)$, G *an arbitrary group,* $A \subset B$ $(1<p<\infty)$. *Then if* A *is complemented in* B *with norm* $\leq K$, A *is invariantly complemented with norm* $\leq K$.

Proof. The assumptions imply that there is a projection from B onto A of norm $\leq K$. So let $\mathcal{P}$ be the set of all projections from B onto A of norm $\leq K$. $\mathcal{P}$ is thus non-empty and convex, and norm-bounded.

Now let X be the space of all bounded linear operators on B, under the weak operator topology. $\mathcal{P}$ is closed in X by A5 of the Appendix; hence $\mathcal{P}$ is compact in X by A6 of the Appendix.

For each $g \in G$, define

$$\mathcal{T}_g : X \to X$$

by

$$\mathcal{T}_g(T) = R_{-g}TR_g \text{ , all } T \in X \text{ .}$$

Then $\mathcal{T}_g(\mathcal{P}) \subset \mathcal{P}$ for each $g \in G$, by the invariance of B and A.

$\{\mathcal{T}_g\}_{g\in G}$ thus forms a commuting family of continuous linear transformations on X, mapping $\mathcal{P}$ into itself. Since $\mathcal{P}$ is convex and compact, there is a fixed-point for the family, by the Markov-Kakutani fixed-point theorem (Theorem 6, p. 456, of [4]).

This means that there is some $P \in \mathcal{P}$ such that

$$\mathcal{T}_g(P) = R_{-g}PR_g = P \text{ , all } g \in G \text{ .}$$

Hence P is a translation-invariant projection onto A, with $\|P\| \leq K$ by the definition of ρ.

Remark 1. It is well known that a bounded linear operator on $L^p(G)$ that commutes with translations is given by a multiplier. I.e., if T is an invariant operator on $L^p(G)$, $1 \leq p \leq 2$, then there is a function h defined on Γ, so that $\widehat{Tf} = h\hat{f}$ a.e., each $f \in L^p$. If T is in addition a projection, then $T^2 = T$, so h must be a characteristic function, i.e. there is some measurable subset E of Γ, cuch that $h = \chi_E$.

Thus let A be an invariant subspace complemented in $L^p(G)$, $1 < p \leq 2$; by Lemma 3.1 there is an invariant projection T from $L^p(G)$ onto A. The above remarks imply that for some measurable subset E of Γ, $\widehat{Tf} = \chi_E \hat{f}$ a.e., for $f \in L^p(G)$. We thus obtain that

$$A = \{f \in L^p(G) : \hat{f}(\tau) = 0 \text{ a.a. } \tau \notin E\},$$

and moreover that χ_E multiplies $\widehat{L^p(G)}$.

Remark 2. Lemma 3.1 holds for non-commutative locally compact amenable groups as well. (For the definition of amenable, see [3]).

Precisely, let G be a non-comutative locally compact group, let μ_R be a fixed right-invariant Haar measure on G, and let A, B be closed right-translation-invariant subspace of $L^p(G, \mu_R)$. I.e., if $g \in G$ let $(R_g f)(x) = f(xg)$, all $x \in G$. Then $R_g f \in A, B$ if $f \in A, B$, respectively.

Suppose that $A \in B$, and A is complemented in B. Then if $1 < p < \infty$ and G is amenable, A is right-invariantly complemented in B. I.e., *there is a bounded linear projection P from B onto A, such that*

P commutes with all right-translations $(R_{g^{-1}} PR_g = P,$ all $g \in G)$.

The proof proceeds along the same lines as the proof given for Lemma 3.1 ; the set $\mathcal{P}$ is analogously defined, and then it is demonstrated that $\mathcal{P}$ is compact in X, X being the space of all bounded linear operators from B onto A under the weak operator topology. The family of operators $\{\mathcal{T}_g\}_{g \in G}$ on X is then given, for each $g \in G$, by

$$T \longrightarrow \mathcal{T}_g(T) = R_{g^{-1}} TR_g, \quad \text{for each } T \in X.$$

Again one shows that $\{\mathcal{T}_g\}_{g \in G}$ is continuous on X and maps $\mathcal{P}$ onto itself.

A projection P in $\mathcal{P}$ that commutes with all right-translations may now be extracted. Indeed, since G is an amenable group, $\{\mathcal{T}_g\}_{g \in G}$ is an amenable group of mappings, so by Theorem 1 of [3], this family has a fixed point in $\mathcal{P}$.

Now we temporarily leave the context of general groups and restrict ourselves to Euclidean space for the next three results.

<u>Theorem</u> 3.2. <u>Let</u> $u \in L^p(\mathbb{R}^n)$, $1 < p < 2$, $u \neq 0$. <u>Then there exists an open set</u> S <u>such that</u> χ_S <u>does not multiply</u> $\hat{u}$.

In order to prove this, we first need

<u>Lemma</u> 3.3. <u>Let</u> $u \in L^p(\mathbb{R}_n)$, $\hat{u}$ <u>compactly supported, and suppose there is a constant</u> C <u>so that</u> χ_S <u>multiplies</u> $\hat{u}$ <u>with norm</u> $\leq C$, <u>for every open set</u> S . <u>Then</u> $u = 0$.

<u>Proof of Lemma</u> 3.3. Fix S an open set, and let $g \in L^p$ be such that $\chi_S \hat{u} = \hat{g}$ a.e.; by hypothesis, $\|g\|_p \leq C\|u\|_p$.

Now note that $g \in L^p(\mathbb{R}^n)$. For $\hat{g} \in L^q$, $\hat{g}$ is compactly supported, whence $\hat{g} \in L^2$, so $g \in L^2$ by Plancherel's theorem.

Now suppose that $v \in L^2 \cap L^q$. Then

$$\left| \int_{\mathbb{R}^n} \hat{g}(x)\hat{v}(x)dx \right| = \left| \int_{\mathbb{R}^n} g(x)v(-x)dx \right|$$

$$\leq \|g\|_p \|v\|_q .$$

(The equality is Parseval's formula; the inequality follows from Hölder's inequality.) Since $\hat{g} = \chi_S \hat{u}$, we thus obtain

$$\left| \int_S \hat{u}(x)\hat{v}(x)dx \right| \leq \|g\|_p \|v\|_q \leq C\|u\|_p \|v\|_q .$$

Now S was an arbitrary open set; since $\hat{u}\hat{v} \in L^1(\mathbb{R}^n)$, we may apply A7 of the Appendix to obtain

$$\int_{\mathbb{R}^n} |\hat{u}(x)\hat{v}(x)| \, dx \leq 4C\|u\|_p \|v\|_q .$$

Now C was independent of v, so this holds for every $v \in L^p \cap L^q$. A lemma of Hörmander's now shows us this is impossible unless $u = 0$.

Lemma 1.2, p. 103, of [9] , implies that if $v \in C_0^{\infty}(\mathbb{R}^n)$, the infinitely differentiable functions with compact support, and if v_t is defined by $\hat{v}_t(\xi) = \hat{v}(\xi)e^{it|\xi|^2}$, (t real), then $\|v_t\|_q \leq C_q |t|^{1/q - 1/2}$, where C_q is independent of t , and $q > 2$. ($q > 2$ here, since $1 < p < 2$.)

So assume $u \neq 0$. Then $\hat{u} \neq 0$, so we may select $v \in C_0^{\infty}(\mathbb{R}^n)$ with

$$0 < \int_{\mathbb{R}^n} |\hat{u}(x)\hat{v}(x)| \, dx .$$

Now

$$|\hat{v}_t| = |\hat{v}| .$$

Thus

$$0 < \int_{\mathbb{R}^n} |\hat{u}(x)\hat{v}(x)|dx$$

$$= \int_{\mathbb{R}^n} |\hat{u}(x)\hat{v}_t(x)| \, dx$$

$$\leq 4C\|u\|_p \|v_t\|_q$$

$$\leq 4C\|u\|_p C_q |t|^{\frac{1}{q}-\frac{1}{2}} \to 0 \text{ as } t \to \infty .$$

This contradiction completes the proof of Lemma 3.3.

We are now able to complete the proof of Theorem 3.2. Let

$$E = \{x \in \mathbb{R}^n : \hat{u}(x) \neq 0\}.$$

(Since $u \neq 0$ is assumed, E is of positive measure.)

Now let $C_1, C_2, \ldots$ be an infinite collection of disjoint cubes, such that $C_i \cap E$ is of positive measure for all i. (C is said to be a cube provided there are real numbers a_i and b_i, $i=1,\ldots n$; such that $C = \{x \in \mathbb{R}^n : a_i < x_i < b_i,\ i = 1,\ldots,n \text{ ; where } x = (x,\ldots,x_n)\}$.)

As is known, there is a constant K (depending only on n and p) such that for all cubes C, χ_C multiplies $\widehat{L^p(\mathbb{R}^n)}$ with norm $\leq K$.

(For example, this follows easily from the theorem in 8.7.11, p. 226, of [15].)

Now define u_n by

$$\hat{u}_n = \chi_{C_n} \hat{u} .$$

We thus have $u_n \neq 0$ for each n. By Lemma 3.3 we may choose an open set $S_n \subset C_n$ such that χ_{S_n} multiplies $\hat{u}_n$ with norm $\geq \frac{n}{\|u_n\|_p}$ (or such that χ_{S_n} does not multiply $\hat{u}_n$ at all; but then $\chi_{S_n \cap C_n}$ would not multiply $\hat{u}$, and we would be done). Now let

$$S = \bigcup_{n=1}^{\infty} S_n .$$

We claim that χ_S does not multiply $\hat{u}$.

Suppose now there were a g such that $g \in L^p$, $\chi_S \hat{u} = \hat{g}$ a.e. Choose n such that

$$K\|g\|_p < n .$$

Finally, define $h \in L^p$ by

$$\hat{h} = \chi_{C_n} \cdot \hat{g} .$$

Then

$$\begin{aligned} \hat{h} = \chi_{C_n} \cdot \chi_S \hat{u} &= \chi_{C_n \cap S} \cdot \hat{u} \\ &= \chi_{S_n} \cdot \chi_{C_n} \hat{u} \\ &= \chi_{S_n} \hat{u}_n \quad \text{a.e.} \end{aligned}$$

Thus

$$n = \frac{n}{\|u_n\|_p} \|u_n\|_p \leq \|h\|_p \leq K\|g\|_p .$$

This contradiction completes the proof.

<u>Corollary</u> 3.4. <u>Let</u> $u \in L^1(\mathbb{R}^n) \cap L^2(\mathbb{R}^n)$, $u \neq 0$. <u>Then there exists an open set</u> S, <u>such that</u> χ_S <u>does not multiply</u> $\hat{u}$ <u>for all</u>

p , $1 < p < 2$, i.e., for all such p , there is no $g \in L^p(\mathbb{R}^n)$ such that $\chi_S \hat{u} = \hat{g}$ a.e.

Proof. Let

$$E = \{x \in \mathbb{R}^n : \hat{u}(x) \neq 0\}.$$

Choose an infinite collection of disjoint cubes $C_1, C_2, \ldots$ such that $C_n \cap E$ is of positive measure. Let u_n be defined by

$$\hat{u}_n = \chi_{C_n} \hat{u} .$$

Thus u_n is a non-zero member of $L^p(\mathbb{R}^n)$, all $1 < p \leq 2$. By Theorem 3.2, for each $n \geq 2$, we may choose an open set $S_n \subset C_n$ such that χ_{S_n} does not multiply $\hat{u}_n$, where u_n is regarded as a member of $L^{1 + 1/n}$.

Then we claim that

$$S = \bigcup_{n=1}^{\infty} S_n$$

is the desired open set.

Now suppose for some $1 < p < 2$ there were a $g \in L^p$ such that $\hat{g} = \chi_S \hat{u}$ a.e. Choose $n \geq 2$ such that $p < 1 + \frac{1}{n}$. Now $\log\|g\|_{1/a}$ is a convex function of a , $0 \leq a \leq 1$, by Lemma 9, p. 524, of [4]; hence $g \in L^r(\mathbb{R}_n)$, all $p \leq r \leq 2$, since $g \in L^2$ by Plancherel's theorem. Hence $g \in L^{1 + 1/n}$, so if we define h by

$$\hat{h} = \chi_{C_n} \hat{g},$$

then $h \in L^{1 + 1/n}$. But

$$\hat{h} = \chi_{C_n} \chi_S \hat{u} = \chi_{S \cap C_n} \hat{u}$$

$$= \chi_{S_n} \chi_{C_n} \hat{u}$$

$$= \chi_{S_n} \hat{u}_n \quad \text{a.e.}$$

Hence χ_{S_n} multiplies $\hat{u}_n$ for $p = 1 + \frac{1}{n}$, a contradiction.

Remark. We discovered Theorem 3.2 and Corollary 3.4 independently of Katznelson, who proved a more general result by defferent methods in [13].

Corollary 3.5. *There exists a bounded open set* S *such that for each* $1 < p < 2$,

$$A_p = \{f \in L^p(\mathbb{R}^n) : \hat{f} = 0 \text{ a.a. } x \notin S\}$$

is uncomplemented in $L^p(\mathbb{R}^n)$.

Proof. Choose S and u satisfying Corollary 3.4, i.e. $u \in L^1 \cap L^2$ and χ_S does not multiply $\hat{u}$ for all $1 < p < 2$. Then A_p is uncomplemented in L^p. For suppose A is complemented (fixing p). Then there is an invariant projection, P, from L^p onto A, by Lemma 3.1. As we pointed out in Remark 1 following Lemma 3.1, there is thus a measurable set E such that $\widehat{Pf} = \chi_E \hat{f}$ a.e., all $f \in L^p$. Choose now an $f \in L^1 \cap L^p$ such that

$$S = \{x \in \mathbb{R}^n : \hat{f}(x) \neq 0\}.$$

f is thus a member of A_p, whence $\widehat{Pf} = \hat{f}$; so we obtain that E differs from S by a set of measure zero. Hence

$$\widehat{Pf} = \chi_S \hat{f} \text{ a.e.},$$

all $f \in L^p$; so χ_S multiplies $\hat{u}$, where u is regarded as a member of L^p, a contradiction.

(Note that in the proof of Corollary 3.4 we could have picked the cubes to be uniformly bounded; i.e., such that there would exist a constant D, so that $x \in C_n$ for some n implies $|x| \leq D$. Then the open set S constructed there would be bounded.)

Q.E.D.

Recall that we are engaged in exhibiting an invariant subspace of A uncomplemented in A, where A is an invariant subspace of $L^p(\mathbb{R}^n)$. (From here until the end of this section p is tacitly assumed to satisfy $1 < p < 2$.) Toward this end we need two lemmas about a particular type of A.

If $u \in L^p(G)$, where G is arbitrary, then A_u is defined to be the closed linear span of all translates of u.

Lemma 3.6. *Let $u \in L^p(G)$ and suppose that $T : A_u \longrightarrow A_u$ is a bounded operator that commutes with translations. Let m be a measurable function so that* $\widehat{Tu} = m\hat{u}$ a. e. with respect to Haar measure on Γ. (Such an m exists; simply define m by $m(x) = \dfrac{\widehat{Tu}(x)}{\hat{u}(x)}$, if $\hat{u}(x) \neq 0$; otherwise let $m(x) = 0$). *Then* $\widehat{Tf} = m\hat{f}$ a. e. *for all* $f \in A_u$.

Proof. Consider an arbitrary linear combination of translates of u; i. e. $f(g) = \sum_{k=1}^{n} a_k \, u(g + t_k)$ for all $g \in G$, where the a_k's are complex constants and the t_k's are in G. Thus

$$f = \sum_{k=1}^{n} a_k R_{t_k} u(g),$$

so

$$\hat{f}(\tau) = \sum_{k=1}^{n} a_k(\tau, t_k)\ \hat{u}(\tau),$$

a. e.; now

$$Tf = \sum_{k=1}^{n} a_k\ T(R_{t_k} u)$$

$$= \sum_{k=1}^{n} a_k\ R_{t_k}\ (T u), \text{ since } T \text{ is invariant.}$$

Hence $\widehat{Tf}(\tau) = \sum_{k=1}^{n} a_k\ \widehat{R_{t_k}(Tu)}(\tau)$

$$= \sum_{k=1}^{n} a_k\ (\tau, t_k)\ \widehat{Tu}(\tau)$$

$$= \sum_{k=1}^{n} a_k\ (\tau, t_k)\ m\hat{u}(\tau)$$

$$= m\ \hat{f}(\tau) \text{ a. e. } \tau \in \Gamma.$$

Now A_u is precisely the closure of all linear combinations of translates of u, by definition. So let $f_n \to f$ in $L^p(G)$, the f_n's being linear combinations of translates of u. Then $\hat{f}_n \to \hat{f}$ in $L^q(G)$ by fact 2. Since T is continuous, $T f_n \to Tf$ in $L^p(G)$ so again $\widehat{Tf}_n \to \widehat{Tf}$ in $L^q(\Gamma)$. We may then choose a subsequence $n_1, n_2, \ldots$ so that $\widehat{Tf}_{n_i} \to \widehat{Tf}$ and $\hat{f}_{n_i} \to \hat{f}$ pointwise a. e.

But then $m\,\hat{f}_{n_i} \to m\hat{f}$ a.e. and $m\,\hat{f}_{n_i} = \widehat{Tf}_{n_i} \to \widehat{Tf}$ a. e., so $\widehat{Tf} = m\hat{f}$ a. e. Since f was an arbitrary element of A_u, this completes the proof of Lemma 3.6.

<u>Lemma</u> 3.7. <u>Let</u> $u \in L^p(G)$, <u>let</u> S <u>be an open subset of</u> Γ, <u>and put</u>

$B = \{g : g \in A_u,\ \hat{g}(\tau) = 0$ for almost all $\tau \notin S\}$.

Then if there is an invariant projection T *from* A_u *onto* B,

$$\widehat{Tg} = \chi_S \hat{g} \quad \text{a. e., } \underline{\text{all}}\ g \in A_u.$$

(i. e. χ_S multiplies $\widehat{A_u}$).

Proof : choose m so that $\widehat{Tu} = m\hat{u}$ a. e., then by Lemma 3.6, $\widehat{Tg} = m\hat{g}$ a.e., for all $g \in A_u$.

Now, let $s \in S$. Then we claim that $\widehat{Tu} = \hat{u}$ a. e. on an open neighborhood of s.

Well, choose $f \in L^1(G)$ so that $\hat{f}$ is identically 1 on an open neighbourhood of s, with $\hat{f}$ being supported in S. Now $f * u \in A_u$ since A_u is closed translation invariant, by fact 1. Thus

$$\widehat{T(f * u)} = m(\widehat{f * u}) = m \hat{f} \hat{u} \quad \text{a. e. and}$$

$$\widehat{f * Tu} = \hat{f} m \hat{u} \quad \text{a. e.,}$$

hence $T(f * u) = f * Tu$ by uniqueness of Fourier transform.

But $f * u \in B$, since $\widehat{f * u} = \hat{f} . \hat{u} = 0$ a.e. off S. Hence $T(f * u) = f * u$.

Thus $\hat{f} . \hat{u} = \hat{f} . \widehat{Tu}$ a. e., so $\widehat{Tu} = \hat{u}$ a. e. in an open neighbourhood of s.

We may thus conclude that $\widehat{Tu} = \hat{u}$ a. e. on S, since there exists an open neighbourhood about each point of S on which this holds.

Finally, $\widehat{Tu} = 0$ a. e. off S, by the definition of B. Hence $\widehat{Tu} = \chi_S \hat{u}$ a. e., so the rest of the assertion follows from Lemma 3.6.

Theorem 3.8. *Let* A *be an invariant non-zero subspace of* $L^p(\mathbb{R}^n)$.

Then there is an invariant subspace of A *that is uncomplemented in* A .

Proof. Let $u \in A$, $u \neq 0$. By Theorem 3.2 there exists an open set S such that χ_S does not multiply $\hat{u}$. Let

$$B = \{f \in A_u : \hat{f}(x) = 0 \text{ a.a. } x \notin S \quad .$$

We claim that B is uncomplemented in A .

For if B were complemented, then by Lemma 3.1 there exists an invariant projection P from A onto B . Since $B \subset A_u$, $P|A_u$ would thus be an invariant projection from A_u onto B . By Lemma 3.7 we would then have that

$$\widehat{Pg} = \chi_S \hat{g} \text{ a.e. },$$

all $g \in A_u$ Hence

$$\widehat{Pu} = \chi_S \hat{u} \text{ a.e.},$$

contradicting the fact that χ_S does not multiply $\hat{u}$.

Q.E.D.

Remark. Lemmas 3.1 and 3.7 actually show that if $u \in L^p(G)$, where G is arbitrary and $1 < p < 2$, and if S is an open subset of Γ such that χ_S does not multiply $\hat{u}$, then if A is any invariant subspace of $L^p(G)$ containing u, there is an invariant subspace of A uncomplemented in A .

Thus, Theorem 3.8 generalizes to any group for which Theorem 3.2 holds.

APPENDIX

The Appendix, with the exception of AO, consists of results which are undoubtedly known, but for which satisfactory references do not seem to be available.

AO. (Completion of the proof of Theorem 1.6.)

Let E be a discrete subset of $\mathbb{R}$ *and a member of the coset-ring of* $\mathbb{R}_d$. *Then there exists a finite set of real numbers* F , *and real numbers* $\tau_1, \ldots, \tau_n$; $\beta_1, \ldots \beta_n$, *such that*

$$E \cup F = \bigcup_{i=1}^{n} (\tau_i \mathbb{Z} + \beta_i) .$$

We first need the following

Lemma. Let G *be a discrete torsion-free abelian group. Let* $\mathcal{A}$ *be all subsets of* G *obtained by intersecting finitely many cosets or complements of cosets. Let* $\mathcal{S}_n$ *be all subsets* S *of* G *having the following property: there exists an infinite subgroup* H *so that*

$$S = \bigcup_{\alpha \in \Lambda} [(x_\alpha + H) \sim F_\alpha] ,$$

where for all α , F_α *is a finite subset of* G *with at most* n *elements,* $x_\alpha \in G$, *and* $(x_{\alpha_1} + H) \cap (x_{\alpha_2} + H) = \emptyset$ if $\alpha_1 \neq \alpha_2$. *Then if* $A \in \mathcal{A}$ *and* A *is infinite,* $A \in \mathcal{S}_n$ *for some* $n \geq 0$.

Proof of Lemma. Since the intersection of two cosets is a coset, every element A of $\mathcal{A}$ may be written as

$$A = C \cap (\sim C_1) \cap (\sim C_2) \ldots \cap (\sim C_n) ,$$

where

$C, C_1, \ldots, C_m$ are cosets of subgroups of G.

Now suppose that A is infinite. Then C is infinite, so $C \in \mathcal{S}_0$. So assume that

$$C \cap (\sim C_1) \cap (\sim C_2) \ldots \cap (\sim C_j) \in \mathcal{S}_j .$$

To show that

$$C \cap (\sim C_1) \cap (\sim C_2) \ldots \cap (\sim C_{j+1}) \in \mathcal{S}_{j+1} ,$$

we need only demonstrate that if $S \in \mathcal{S}_j$ and if D is a coset of G, then $(S \sim D) \in \mathcal{S}_{j+1}$ (provided $(S \sim D) \neq \emptyset$). The proof will then be complete, by mathematical induction.

So, assuming $S \in \mathcal{S}_j$, let

$$S = \bigcup_{\alpha \in \Lambda} [(x_\alpha + H) \sim F_\alpha] ,$$

where H is an infinite subgroup and each F_α is a finite subset of G having at most j elements. Let

$$D = y + Q ,$$

where $y \in G$, Q is a subgroup of G.

Case (a). $H \cap Q = \{0\}$.

Then for each $\alpha \in \Lambda$, $(x_\alpha + H) \cap (y+Q)$ is either empty, or consists of one point. So define

$$F_\alpha^* = F_\alpha \cup [(x_\alpha + H) \cap (y+Q)] ;$$

then F_α^* has at most j+1 elements for all α, and

$$S \sim D = \bigcup_{\alpha \in \Lambda} [(x_\alpha + H) \sim F_\alpha^*] ;$$

hence $S \sim D \in \mathcal{S}_{j+1}$.

Case (b). $H \cap Q \neq \{0\}$.

Let $H^* = H \cap Q$. Since G is torsion-free, H^* is an infinite subgroup. Now fix $\alpha \in \Lambda$. If $(x_\alpha + H) \cap (\sim(y+Q))$ is non-empty, it must be a union of cosets of H^*. Indeed, for each $z \in G$, $(x_\alpha + H) \cap (z+Q)$ is either empty or a coset of H^*, and

$$\sim D = \sim(y+Q) = \bigcup_{z \notin y+Q} (z+Q) .$$

It follows immediately that

$$[x_\alpha + H] \bigcap [\sim(y+Q)] \cap [\sim F_\alpha] = [(x_\alpha + H) \sim F_\alpha] \cap [\sim D] \in \mathcal{S}_j ,$$

since this set differs by the finite set F_α from a union of cosets of H^*, and F_α has at most j elements.

Now let

$$\Lambda^* = \{\alpha \in \Lambda : (x_\alpha + H) \cap (\sim D) \neq \emptyset\} .$$

Λ^* is non-empty since $S \sim D$ is assumed non-empty. Then

$$S \sim D = \bigcup_{\alpha \in \Lambda^*} [(x_\alpha + H) \cap (\sim F_\alpha) \cap (\sim D)]$$

is a member of $\mathcal{S}_j$, since a union of members of $\mathcal{S}_j$, all corresponded to the same subgroup of G, belongs to $\mathcal{S}_j$ (the subgroup in this case being H^*).

Finally $\mathcal{S}_j \subset \mathcal{S}_{j+1}$, so the proof of the lemma is complete.

We are now ready to prove AO, and hence complete the proof of Theorem 1.6.

Let $\mathcal{A}$ be the members of $\mathcal{R}$ obtained by intersecting finitely many cosets or complements of cosets. (Recall that $\mathcal{R}$ is the coset-ring of the discrete real line.) $\mathcal{R}$ consists of all finite unions of members of $\mathcal{A}$, so there are sets $S_i \in \mathcal{A}$ and an integer m such that $E = \bigcup_{i=1}^{m} S_i$. If E is infinite, then one of the S_i's must be infinite; moreover, all of the S_i's must be discrete, since E is.

Now if S is an infinite member of $\mathcal{A}$, then by our lemma, there exists an integer n and an infinite subgroup H, such that

$$S = \bigcup_{\alpha \in \Lambda} [(x_\alpha + H) \sim F_\alpha] ,$$

where F_α is a finite set with at most n elements, for all α, and the x_α's are real numbers with $(x_{\alpha_1} + H) \cap (x_{\alpha_2} + H) = \emptyset$ if $\alpha_1 \neq \alpha_2$.

If S is in addition a discrete subset of $\mathbb{R}$, then H must be discrete and Λ must be finite. For H is an infinite subgroup of $\mathbb{R}$; if H is not discrete, then H is dense in $\mathbb{R}$; hence for a fixed α, so is $(x_\alpha + H) \sim F_\alpha$.

Thus, since H is discrete and infinite, there exists a real τ, so that $H = \tau \mathbb{Z}$. Now for each $\alpha \in \Lambda$,

$$x_\alpha + \tau \mathbb{Z} = x'_\alpha + \tau \mathbb{Z}, \text{ where } 0 \leq x'_\alpha < \tau .$$

So we may write

$$S = \bigcup_{\alpha \in \Lambda} [(x'_\alpha + \tau \mathbb{Z}) \sim F_\alpha]$$

where $x'_{\alpha_1} \neq x'_{\alpha_2}$ if $\alpha_1 \neq \alpha_2$. Since S is discrete, S contains only finitely many elements, say k of them, in the interval $[0, (n+1)\tau]$. (Recall that F_α has at most n elements, all α.) Then Λ has at most k elements, since $x'_{\alpha_1} + j\tau \neq x'_{\alpha_2} + \ell\tau$ if $\alpha_1 \neq \alpha_2$, and j, ℓ are arbitrary integers.

Thus if S is an infinite discrete member of $\mathcal{A}$,

$$S = \bigcup_{i=1}^{k} [(x_i + \tau \mathbb{Z}) \sim F_i]$$

for some real numbers x_i, finite sets F_i. Or, letting $F = \bigcup_{i=1}^{k} F_i$,

$$S \cup F = \bigcup_{i=1}^{k} (x_i + \tau \mathbb{Z}) .$$

If S is a finite member of $\mathcal{Q}$, we may write $S = \bigcup_{s \in S} (s + 0 \cdot \mathbb{Z})$. Since E, our discrete member of $\mathcal{R}$, is a finite union of such S's, A0 is established.

<u>A1</u>. <u>Let</u> $(X, \mathcal{S}, \mu)$ <u>be a finite measure space</u>, $1 \leq p < 2$, <u>and</u> A a <u>closed subspace of</u> $L^p(X)$ <u>with</u> $A \subset L^2(X)$. <u>Then the</u> L^p <u>and</u> L^2 <u>norms on</u> A <u>are equivalent</u>.

<u>Proof</u>. By Hölder's inequality

$$\int |f|^p \leq \left(\int |f|^2\right)^{p/2} (\mu(X))^{1-p/2} .$$

Hence

$$\|f\|_p \leq C \|f\|_2 ,$$

all $f \in L^p(X)$, where $C = (\mu(X))^{1/p - 1/2}$. So it suffices to show that there is a K, $\|f\|_2 \leq K \|F\|_p$, all $f \in A$.

Suppose $f_n \in A$, $f_n \to f$ in L^p and $f_n \to g$ in L^2. We may choose a subsequence f_{n_i}, so that $f_{n_i} \to f$ a.e., and $f_{n_i} \to g$ a.e. Thus $f = g$ a.e. Hence, by the closed graph theorem, the identity map from A with L^p norm onto A with L^2 norm is continuous.

Q.E.D.

<u>A2</u>. <u>Let</u> $(X, \mathcal{S}, \mu)$ <u>a finite measure space</u>, $1 < r < \infty$, <u>and</u> A a <u>closed subspace of</u> $L^r(X)$. <u>Suppose there is a constant</u> C, <u>and an</u> s, $1 \leq s < r$, <u>so that</u>

$$\|f\|_r \leq C \|f\|_s , \text{ all } f \in A .$$

Then the L^t norms are all equivalent, $1 \leq t \leq r$.

Proof. Suppose, for the sake of convenience, that $\mu(X) = 1$. It suffices to show that there is a K, such that $\|f\|_s \le K\|f\|_1$, for all $f \in A$. For if $1 \le t_1 \le t_2 \le r$, then $\|f\|_{t_1} \le \|f\|_{t_2}$ for all $f \in L^r(X)$. Indeed, by Hölder's inequality, we have

$$\int |f|^{t_1} \le \left(\int |f|^{t_2}\right)^{t_1/t_2} \left(\int (1)^k d\mu(x)\right)^{\frac{1}{k}}$$

where $\frac{1}{k} = 1 - t_1/t_2$. Thus, if $f \in A$, and t_1, t_2 are two points in $[1, r]$,

$$\|f\|_{t_1} \le \|f\|_r \le C\|f\|_s \le KC\|f\|_1 \le KC\|f\|_{t_2}.$$

Thus the proof will be complete if we can establish the existence of K.

Solving the equations

$$t_1 + t_2 = 1$$

$$t_1 + rt_2 = s,$$

we obtain $t_2 = \frac{s-1}{r-1}$. Thus

$$0 < t_i < 1, \quad i = 1, 2.$$

Suppose $f \in A$, $f \ne 0$.

$$\begin{aligned} \int |f|^s &= \int |f|^{t_1} |f|^{rt_2} \\ &\le \left(\int |f|\right)^{t_1} \left(\int |f|^r\right)^{t_2} \end{aligned}$$

by Hölder's inequality. But

$$\left(\int |f|^r\right)^{t_2} \le C^{t_2 r} \left(\int |f|^s\right)^{t_2 r/s}$$

by assumption. Hence

$$\left(\int |f|^s\right)^{\frac{1}{t_1} - \frac{t_2 r}{t_1 s}} \le C^{t_2 r/t_1} \int |f|.$$

$$\frac{1}{t_1} - \frac{t_2 r}{t_1 s} = \frac{s - r t_2}{t_1 s} = \frac{1}{s} .$$

Thus $K = C^{t_2 r/t_1}$ works.

A3. Let $(X, \mathcal{S}, \mu)$ be a finite measure space, $2 < p < \infty$, and A a closed subspace of $L^p(X)$ with the L^2 norm equivalent to the L^p norm on A. Then A is complemented in $L^p(X)$.

Proof. By Holder's inequality, if $f \in L^p(X)$,

$$\int_X |f|^2 d\mu \leq \left(\int |f|^p\right)^{\frac{2}{p}} (\mu(X))^{1-\frac{2}{p}} .$$

Thus, if $f \in L^p(X)$, $\|f\|_2 \leq K_1 \|f\|_p$, where $K_1 = (\mu(X))^{\frac{1}{2}-\frac{1}{r}}$.

Now let P be the orthogonal projection from $L^2(X)$ onto A, where A is regarded as a closed subspace of $L^2(X)$. Then P is actually a bounded operator on $L^p(X)$; for, if $f \in L^p(X)$,

$$\|Pf\|_2 \leq \|f\|_2 \leq K_1 \|f\|_p .$$

But by assumption there is a constant K_2 so that for all $g \in A$,

$$\|g\|_p \leq K_2 \|g\|_2 .$$

Hence

$$\|Pf\|_p \leq K_2 \|Pf\|_2 \leq K_1 K_2 \|f\|_p .$$

Thus P is a bounded projection from $L^p(X)$ onto A, so A is complemented in $L^p(X)$.

A4. Let B a Banach space, let X be the bounded operators on B under the strong operator topology. Let ρ be a compact subset of X. Then the closed convex hull of ρ is a compact subset of X.

Proof. For each $f \in B$, let B_f be a copy of B. Let

$$Y = \prod_{f \in B} B_f .$$

Now Y consists of all functions from B to B ; hence $X \subset Y$. Moreover, if y_α is a net in Y , then $y_\alpha \to y$ if and only if $y_\alpha(f) \to y(f)$ in B_f , each $f \in B$. Hence if y_α , $y \in X$, then $y_\alpha \to y$ in X if and only if $y_\alpha \to y$ in Y.

Thus X is a subspace of Y, and the relative Y-topology on X is the same as the strong operator topology. Moreover, if A is a norm-bounded subset of X , the closure of A in X is actually a closed subset of Y .

For let $y_\alpha \in A$, $y_\alpha \to y$, $y \in Y$. y is a linear operator on B since the y_α's are. $\|y_\alpha\| \leq K$, hence $\|y_\alpha(f)\|_B \leq K\|f\|_B$, all $f \in B$, all α . y is thus a bounded linear operator on B , so $y \in X$.

Now let $\pi_f : Y \to B_f$ be defined by

$$\pi_f(y) = y(f) , \text{ all } y \in Y .$$

π_f is continuous by definition of the product topology; hence $\pi_f(P)$ is a compact subset of B , each $f \in B$. Now

$$P \subset \prod_{f \in B} \pi_f(P) .$$

Hence

$$\text{convex hull } P \subset \prod_{f \in B} S_f ,$$

where

$$S_f = \text{closed convex hull of } \pi_f(P) \text{ in } B .$$

But the closed convex hull of a compact subset of a Banach space is compact (Theorem 6, p. 416, of [4]). Hence $\prod_{f \in B} S_f$ is compact by the Tychonoff theorem, since S_f is compact for each f .

Since $\mathcal{P}$ is compact, one can show by the uniform boundedness theorem that $\mathcal{P}$ is norm-bounded.

Finally, the convex hull of $\mathcal{P}$ is again norm-bounded, so the closure of the convex hull of $\mathcal{P}$ in X is already a closed subset of Y , and is hence compact, being a closed subset of the compact set $\prod_{f \in B} S_f$.

A5. Let B a Banach space, A a closed subspace of B , and $\mathcal{P}$ a subset of the bounded projections from B onto A .

Then (a): the convex hull of $\mathcal{P}$,

(b): the closed convex hull of $\mathcal{P}$ in the strong operator topology,

and (c): the closed convex hull of $\mathcal{P}$ in the weak operator topology; all consist of bounded projections from B onto A .

Proof. Let $0 \leq \lambda_i \leq 1$, $1 \leq i \leq n$, $\sum_{i=1}^{n} \lambda_i = 1$, and $P_1, \ldots, P_n \in \mathcal{P}$.

If $F \in A$, then $P_i(f) = f$; hence

$$\sum_{i=1}^{n} \lambda_i P_i f = \sum_{i=1}^{n} \lambda_i f = f .$$

If $f \in B$, then $P_i f \in A$; hence $\sum_{i=1}^{n} \lambda_i P_i(f) \in A$, since A is a linear space.

Hence $\sum_{i=1}^{n} \lambda_i P_i$ is a bounded projection from B onto A , so (a) is proved.

Since the strong operator topology is stronger than the weak operator topology, (c) $\Longrightarrow$ (b). So it suffices to prove (c) .

Suppose T_α is a net in the convex hull of $\mathcal{P}$, $T_\alpha \to T$ in the weak operator topology. $T_\alpha(f) = f$, all $f \in A$, so $T(f) = f$. Finally, the Hahn-Banach theorem implies that A is weakly closed, since A is a norm-closed subspace of B . Thus, since $T_\alpha(f) \in A$, all $f \in B$, all α ,

$$\lim_\alpha T_\alpha(f) = Tf \in A ,$$

since Tf is thus a weak limit of members of A . Hence T is a projection onto A .

<u>A6. Let B be a reflexive Banach space. Then a norm-bounded subset of the bounded operators on B is relatively compact in the weak operator topology.</u>

<u>Proof</u>. Let X denote the linear space B under the weak topology, i.e., $x_\alpha \to x$ if and only if $f(x_\alpha) \to f(x)$, all $f \in B^*$.

For each $f \in B^*$, let X_f be a copy of X .

Let Y denote the space of bounded operators on B under the weak operator topology. Then Y is a subspace of $\prod_{f \in B^*} X_f$, and the weak operator topology on Y is the relative product topology.

Now if $\mathcal{P}$ is a norm-bounded subset of X , then $\pi_f(\mathcal{P})$ is a norm-bounded subset of X . (π_f is the natural projection from

$\prod_{f\in B^*} X_f$, onto X_f, defined as in A4.) Hence $\overline{\pi_f(\mathcal{P})}$ is compact in X_f by Corollary 8, p. 425, of [4]. Thus $\prod_{f\in B^*} \overline{\pi_f(\mathcal{P})}$ is compact by the Tychonoff theorem. Since

$$\bar{\mathcal{P}} \subset \prod_{f\in B^*} \overline{\pi_f(\mathcal{P})}$$

$\bar{\mathcal{P}}$ is compact (where $\bar{\mathcal{P}}$ denotes the product closure of $\mathcal{P}$).

Finally, as in A4, since $\mathcal{P}$ is norm-bounded, the closure of $\mathcal{P}$ in Y is a closed subset of $\prod_{f\in B^*} X_f$, so we have that $\bar{\mathcal{P}} \subset Y$.

Hence $\mathcal{P}$ is a relatively compact subset of Y.

A7. Let $f \in L^1(\mathbb{R}^n)$ and suppose there is a constant K such that $\left|\int_S f\right| \le K$ for every open set S. Then $\int_{\mathbb{R}^n} |f| \le 4K$.

Proof. Choose measurable sets E_1 and E_2 such that

$$\left|\int_{E_1} \operatorname{Re} f\right| \ge \frac{1}{2} \int_{\mathbb{R}^n} |\operatorname{Re} f|$$

and

$$\left|\int_{E_2} \operatorname{Im} f\right| \ge \frac{1}{2} \int_{\mathbb{R}^n} |\operatorname{Im} f| .$$

Then

$$\left|\int_{E_1} f\right| + \left|\int_{E_2} f\right| \ge \left|\int_{E_1} \operatorname{Re} f\right| + \left|\int_{E_2} \operatorname{Im} f\right|$$

$$\ge \frac{1}{2} \int_{\mathbb{R}^n} |\operatorname{Re} f| + |\operatorname{Im} f|$$

$$\geq \frac{1}{2} \int_{\mathbb{R}^n} |f| .$$

Letting $E = E_i$, where

$$\left|\int_{E_i} f\right| = \max\left\{\left|\int_{E_1} f\right| , \left|\int_{E_2} f\right|\right\} ,$$

we thus have

$$\left|\int_E f\right| \geq \frac{1}{4} \int_{\mathbb{R}^n} |f| .$$

Now let $\epsilon > 0$ be given. Since fdx is absolutely continuous with respect to Lebesgue measure, denoted by λ, we may choose a $\delta > 0$, so that

$$\lambda(F) < \delta \Longrightarrow \int_F |f|\, dx < \epsilon .$$

Now choose S an open set with $E \subset S$ and $\lambda(S \sim E) < \delta$. Then

$$K \geq \left|\int_S f\right| = \left|\int_E f + \int_{S\sim E} f\right|$$

$$\geq \left|\int_E f\right| - \left|\int_{S\sim E} f\right|$$

$$\geq \frac{1}{4} \int_{\mathbb{R}^n} |f| - \epsilon .$$

Hence

$$\int_{\mathbb{R}^n} |f| \leq 4K + 4\epsilon .$$

Since this holds for every $\epsilon > 0$, we have

$$\int_{\mathbb{R}^n} |f| \leq 4K .$$

Q.E.D.

BIBLIOGRAPHY

[1] Banach, S., Theorie des Operations Lineaires, Monografje Matematyczne, Warsaw, 1932.

[2] Day, M.M., Normed Linear Spaces, Ergebnisse der Mathematik und ihrer Grenzgebiete, n. F., Heft 21, Springer, Berlin, 1958.

[3] Day, M.M., "Fixed-Point Theorems for Compact Convex Sets", Illinois J. Math., 5, (1961) 585-590.

[4] Dunford, N., and Schwartz, J. T., Linear Operators. Part I : General Theory, Interscience Pubs., New York, 1958.

[5] Eberlein, W.F., "Characterizations of Fourier-Stieltjes Transforms", Duke Math. J., 22 (1955), 465-468.

[6] Figa-Talamanca, A., "Multipliers of p-Integrable Functions," Bull. Amer. Math. Soc., 70 (1964), 666-669.

[7] Glicksberg, I., "Some Uncomplemented Function Algebras," Trans. Amer. Math. Soc., 111 (1964), 121-137.

[8] Hewitt, E., "A Survey of Abstract Harmonic Analysis. Some Aspects of Analysis and Probability," pp. 105-168 in Surveys in Applied Mathematics, vol. 4, John Wiley and Sons, Inc., New York, 1958.

[9] Hörmander, L., "Estimates for Translation Invariant Operators in L^p Spaces," Acta Math., 104 (1960), 93-140.

[10] Isbell, J.R., and Semadini, Z., "Projection Constants and Spaces of Continuous Functions," Trans. Amer. Math. Soc., 107 (1963), 38-48.

[11] Itô, T., and Amemiya, I., "A simple proof of the Theorem of P.J. Cohen," Bull. Amer. Math. Soc., 70 (1964), 774-776.

[12] Kadec, M.I., and Pełczynski, A., "Bases, Lacunary Sequences, and Complemented Subspaces in the Spaces L_p," Studia Math., 21 (1962), 161-176.

[13] Katznelson, Y., "Sets of Uniqueness for Some Classes of Trigonometrical Series," Bull. Amer. Math. Soc., 70 (1964), 722-723.

[14] Pełczynski, A., "On Simultaneous Extension of Continuous Functions. (A Generalization of Theorems of Rudin-Carleson and Bishop)," Studia Math., 24 (1964), 285-304.

[15] Rudin, W., Fourier Analysis on Groups, Interscience Pubs., New York, 1962.

[16] Rudin, W., "Projections on Invariant Subspaces," Proc. Amer. Math. Soc., 13 (1962), 429-432.

[17] Rudin, W., "Trigonometric Series with Gaps," J. Math. Mech., 9 (1960), 203-227.

[18] Šilov, G.E., "Homogeneous Rings of Functions," Amer. Math. Soc. Translation No. 92, translated from Uspehi Matem. Nauk (N.S.), 6 (1951), 91-137. Amer. Math. Soc., Providence, R.I., 1953.

[19] von Neumann, J., "Zur allgemeinen Theorie des Masses," Fund. Math., 13 (1929), 73-116.

[20] Zygmund, A., Trigonometric Series, Vol. I, 2nd edition, Cambridge University Press, Cambridge, England, 1959.